Collins

The Shangh Maths Project

For the English National Curriculum

Practice Book 2B

Series Editor: Professor Lianghuo Fan

UK Curriculum Consultant: Paul Broadbent

William Collins' dream of knowledge for all began with the publication of his first book in 1819.

A self-educated mill worker, he not only enriched millions of lives, but also founded a flourishing publishing house. Today, staying true to this spirit, Collins books are packed with inspiration, innovation and practical expertise. They place you at the centre of a world of possibility and give you exactly what you need to explore it.

Collins. Freedom to teach.

Published by Collins, an imprint of HarperCollins*Publishers*
The News Building
1 London Bridge Street
London
SE1 9GF

© HarperCollins*Publishers* Limited 2017
© Professor Lianghuo Fan 2017
© East China Normal University Press Ltd. 2017

Browse the complete Collins catalogue at
www.collins.co.uk

10 9 8 7 6 5 4 3 2 1

978-0-00-822610-7

Translated by Professor Lianghuo Fan, Adapted by Professor Lianghuo Fan.

British Library Cataloguing in Publication Data

A catalogue record for this publication is available from the British Library.

Series Editor: Professor Lianghuo Fan
UK Curriculum Consultant: Paul Broadbent
Publishing Manager: Fiona McGlade
In-house Editor: Nina Smith
In-house Editorial Assistant: August Stevens
Project Manager: Emily Hooton
Copy Editor: Karen Williams
Proofreaders: Cassandra Fox and Gerard Delaney
Cover design: Kevin Robbins and East China University Press Ltd
Cover artwork: Daniela Geremia
Internal design: 2Hoots Publishing Services Ltd
Typesetting: Ken Vail Graphic Design Ltd
Illustrations: QBS
Production: Rachel Weaver

Printed and bound by CPI Group (UK) Ltd, Croydon, CR0 4YY

The Shanghai Maths Project (for the English National Curriculum) is a collaborative effort between HarperCollins, East China Normal University Press Ltd. and Professor Lianghuo Fan and his team. Based on the latest edition of the award-winning series of learning resource books, One Lesson, One Exercise, by East China Normal University Press Ltd. in Chinese, the series of Practice Books is published by HarperCollins after adaptation following the English National Curriculum.

Practice Book Year 2B is translated and developed by Professor Lianghuo Fan with assistance of Ellen Chen, Ming Ni, Huiping Xu and Dr. Lionel Pereira-Mendoza, with Paul Broadbent as UK Curriculum Consultant.

Contents

Chapter 5 Multiplication and division (I)

5.1 Introduction to multiplication (1)

 Learning objective Multiply by grouping and repeated addition

 Basic questions

1 Look at the pictures and fill in the answers.

3 twos = ☐ + ☐ + ☐ = ☐

4 threes = ☐ + ☐ + ☐ + ☐ = ☐

5 fours = ☐ + ☐ + ☐ + ☐ + ☐ = ☐

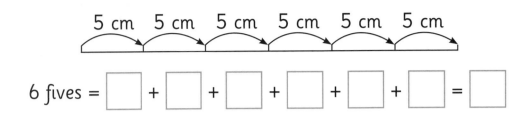

6 fives = ☐ + ☐ + ☐ + ☐ + ☐ + ☐ = ☐

2 Group the objects and fill in the boxes. Then write the addition sentences.

(a)

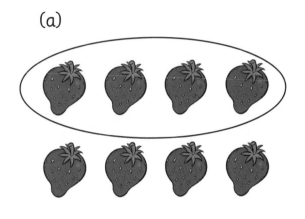

2 groups of 4 = ☐

4 groups of 3 = ☐

Addition sentence:

Addition sentence:

_____

_____

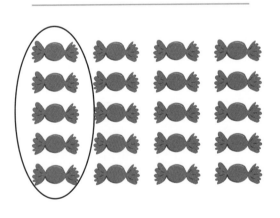

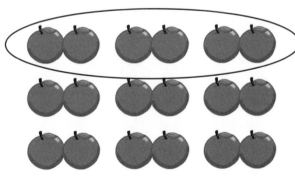

☐ groups of ☐ = ☐

☐ groups of ☐ = ☐

Addition sentence:

Addition sentence:

_____

_____

(b) Fill in the boxes with suitable numbers.

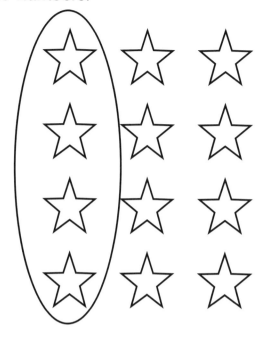

☐ groups of 3 = ☐   ☐ groups of ☐ = ☐

☐ groups of ☐ = ☐

3 Equal groups. Look at this picture of a classroom and then fill in the boxes.

2 pupils share 1 desk. There are ☐ desks of 2 in the classroom.

There are ☐ pupils in each group. There are ☐ groups of ☐ in the class.

There are ☐ pencils in each pot. There are ☐ pots of ☐ in the class.

Challenge and extension question

4 Think carefully and then write the answers.

◯ + △ = ☐10

◯ + ◯ + △ + △ + △ = ☐23

◯ = ☐

△ = ☐

4

5.2 Introduction to multiplication (2)

 Learning objective Multiply by grouping and repeated addition

 Basic questions

1 Look at the pictures and fill in the boxes. Then write the addition and multiplication sentences.

(a)

☐ groups of ☐ = ☐

Addition sentence:

_____

Multiplication sentence:

_____

(b)

☐ groups of ☐ = ☐

Addition sentence:

_____

Multiplication sentence:

_____

☐ groups of ☐ = ☐

Addition sentence:

_____

Multiplication sentence:

_____

☐ groups of ☐ = ☐

Addition sentence:

_____

Multiplication sentence:

_____

2 Group the objects, fill in the boxes and then complete the number sentences.

(a)

☆ ☆ ☆ ☆ ☆   ☆ ☆ ☆ ☆ ☆
☆ ☆ ☆ ☆ ☆   ☆ ☆ ☆ ☆ ☆
☆ ☆ ☆ ☆ ☆   ☆ ☆ ☆ ☆ ☆
☆ ☆ ☆ ☆ ☆   ☆ ☆ ☆ ☆ ☆

4 groups of 5 are ☐.   5 groups of 4 are ☐.

Multiplication sentence: Multiplication sentence:

_____   _____

$$☐ \times ☐ = ☐ \times ☐$$

The two factors are 4 and ☐.

The product is ☐.

(b)

○ ○ ○ ○ ○ ○   ○ ○ ○ ○ ○ ○
○ ○ ○ ○ ○ ○   ○ ○ ○ ○ ○ ○
○ ○ ○ ○ ○ ○   ○ ○ ○ ○ ○ ○

☐ groups of ☐ are ☐.   ☐ groups of ☐ are ☐.

Multiplication sentence: Multiplication sentence:

_____   _____

$$☐ \times ☐ = ☐ \times ☐$$

The two factors are ☐ and 3. The product is ☐.

3 Change each addition sentence into a multiplication sentence.

(a) 2 + 2 + 2 = ☐   ☐ × ☐ = ☐

(b) 4 + 4 + 4 = ☐   ☐ × ☐ = ☐

(c) 8 + 8 + 8 + 8 = ☐   ☐ × ☐ = ☐

(d) 6 + 6 + 6 + 6 + 6 = ☐   ☐ × ☐ = ☐

4 Fill in the ☐ with correct numbers and the ◯ with >, < or =.

(a) 4 × 7 = 7 + 7 + 7 + 7 = ☐

7 × 4 = 4 + 4 + 4 + 4 + 4 + 4 + 4 = ☐

4 × 7 ◯ 7 × 4

(b) 3 × 5 = 5 + 5 + 5 = ☐

5 × 3 = 3 + 3 + 3 + 3 + 3 = ☐

3 × 5 ◯ 5 × 3

(c) 2 × 6 = ☐ × ☐

(d) 5 × 7 = ☐ × ☐

(e) 3 × 8 = ☐ × ☐

(f) 4 × 9 = ☐ × ☐

Challenge and extension question

5 Think carefully. Then write the answer in the box.

If ◇ + ◇ + ◇ = 27,

☆ + ☆ + ☆ + ☆ = 40,

● + ● + ● + ● + ● = 25,

work out ◇ + ☆ + ● = ☐.

5.3 Writing multiplication sentences with pictures (1)

 Learning objective Multiply by grouping and repeated addition

 Basic questions

1 Calculate mentally.

3 + 3 + 3 + 3 + 3 = ☐   8 + 8 + 8 + 8 = ☐   6 + 6 + 6 = ☐

5 + 5 + 5 + 5 + 5 = ☐   4 + 4 + 4 + 4 = ☐   9 + 9 + 9 = ☐

2 Look at the pictures and write the number sentences.

(a)

How many chicks are there altogether?

Addition sentence: _____

Multiplication sentence: _____

(b)

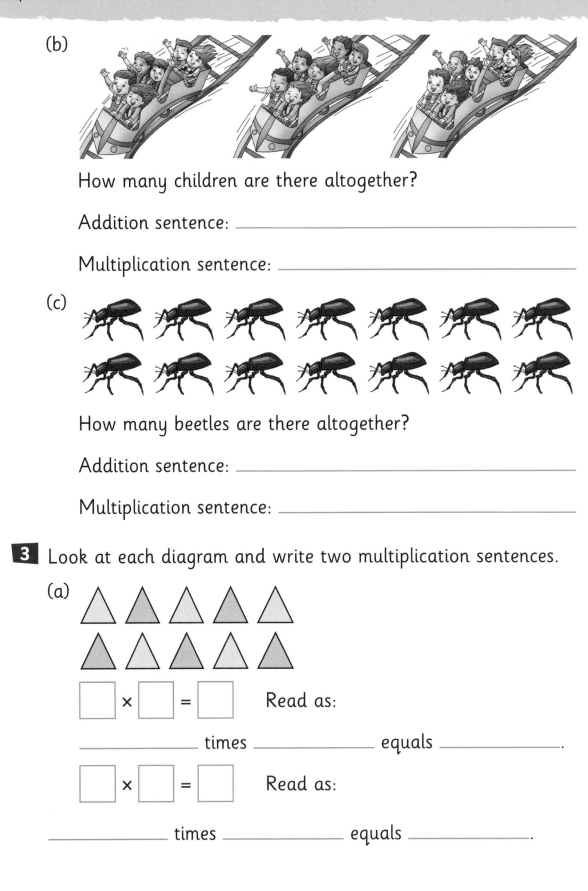

How many children are there altogether?

Addition sentence: _____

Multiplication sentence: _____

(c)

How many beetles are there altogether?

Addition sentence: _____

Multiplication sentence: _____

3 Look at each diagram and write two multiplication sentences.

(a)

☐ × ☐ = ☐   Read as:

_____ times _____ equals _____.

☐ × ☐ = ☐   Read as:

_____ times _____ equals _____.

(b)

☆★☆★☆☆
★☆★☆★☆
☆★☆★☆★

☐ × ☐ = ☐   Read as:

_____ times _____ equals _____.

☐ × ☐ = ☐   Read as:

_____ times _____ equals _____.

(c)

☐ × ☐ = ☐   Read as:

_____ times _____ equals _____.

☐ × ☐ = ☐   Read as:

_____ times _____ equals _____.

Challenge and extension question

4   Think carefully and then fill in the answer boxes.

☆☆☆   ☆☆☆   ☆☆☆   ☆☆

☐ × ☐ + ☐ = ☐

☐ × ☐ – ☐ = ☐

5.4 Writing multiplication sentences with pictures (2)

 Learning objective Use an array to show the commutative nature of multiplication

 Basic questions

1 Look at the pictures and write your answers underneath.

There are ☐ flowers in each vase and there are ☐ vases.

There are ☐ flowers altogether.

It means ☐ groups of ☐.

Multiplication sentence: _____

2 Write two multiplication sentences for each diagram.

(a)   (b)

_____   _____

_____   _____

(c)  (d)

_____

_____

(e)  (f)

_____

_____

(g) (h)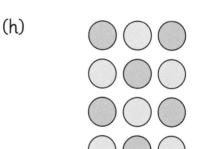

_____

_____

(i)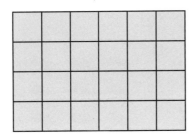

(j)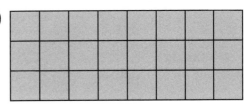

_____   _____

_____   _____

Challenge and extension question

3 Think carefully. How can you use multiplication to get the same answer?

(a) $4 + 4 + 4 + 4 + 3 =$ ☐

_____

(b) $4 + 4 + 4 + 4 + 5 =$ ☐

_____

(c) $4 + 4 + 4 + 4 + 5 + 3 =$ ☐

_____

5.5 Writing multiplication sentences with pictures (3)

Learning objective Use an array to show the commutative nature of multiplication

Basic questions

1 Fill in the boxes with suitable numbers.

3 × 2 = ☐ × ☐   ☐ × ☐ = 2 × 8

☐ × 6 = ☐ × 5   4 × 6 = ☐ × ☐

☐ × ☐ = 5 × 7   4 × ☐ = 8 × ☐

5 × 7 = ☐ × ☐   ☐ × ☐ = 6 × 8

☐ × ☐ = ☐ × ☐

2 Look at the pictures and then complete the multiplication calculations.

(a)

___ × ___ = ___ × ___ = ___

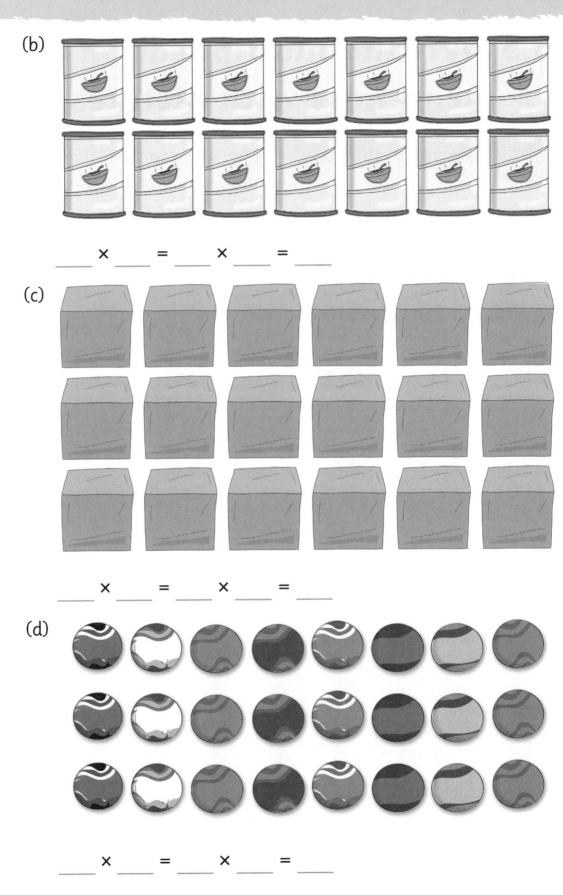

(b)

____ × ____ = ____ × ____ = ____

(c)

____ × ____ = ____ × ____ = ____

(d)

____ × ____ = ____ × ____ = ____

(e)

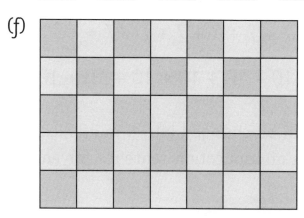

_____ × _____ = _____ × _____ = _____

(f)

_____ × _____ = _____ × _____ = _____

3 Draw a diagram for each multiplication below.
Then change the order of the factors to write another
multiplication sentence.

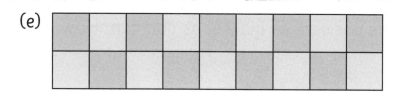

7 × 3

2 × 6

Diagram: Diagram:

Multiplication sentence: Multiplication sentence:

_____   _____

4 Calculate the answers as quickly as you can. How will you do it?

(a) 4 + 4 + 4 + 4 + 4 + 4 + 4 = ☐

(b) 5 + 5 + 5 + 5 + 5 + 5 + 5 + 5 + 5 = ☐

(c) 7 + 7 + 7 + 7 + 7 + 7 + 7 + 7 + 7 + 7 = ☐

(d) 10 + 10 + 10 + 10 + 10 + 10 + 10 + 10 + 10 + 10 = ☐

5 Pick the fruit that can be changed into a multiplication sentence and write a multiplication sentence for each.

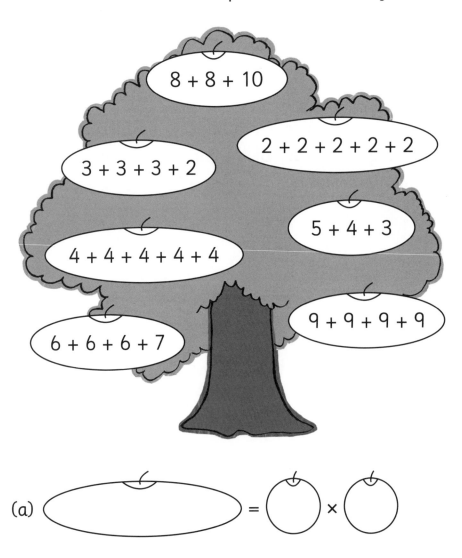

(a)

(b)

(c)

6 Think carefully and then fill in the boxes.

◇ + ◇ = △ + △ + △ = ☆ + ☆ + ☆ + ☆ = 12

◇ = ☐   △ = ☐   ☆ = ☐

5.6 Times (1)

Learning objective Use grouping and repeated addition to model multiplication and the multiplication (×) sign

Basic questions

1 Look at the pictures, fill in the boxes and complete the number sentences.

(a)

There are 5 apples in 1 group. There are ☐ groups, so there are ☐ groups of ☐.

We can also say that the number of apples is ☐ times ☐.
Multiplication sentence:

_____ or _____

(b)

There are 3 sticks in 1 group. There are ☐ groups, so there are ☐ groups of ☐.

We can also say that the number of sticks is ☐ times ☐.
Multiplication sentence:

_____ or _____

2 Draw ◯ so that the number of ◯ is 3 times the number of △.

(a)  △ △ △ _____

(b)  △ △ △ △ △ _____

3 Write the correct numbers into the boxes.

(a)

4 + 4 = ☐

☐ fours

☐ times two

4 × 2 = ☐

2 × 4 = ☐

(b)

5 + 5 + 5 + 5 = ☐

☐ fives

☐ times ☐

☐ × ☐ = ☐

☐ × ☐ = ☐

(c)

10 + 10 + 10 + 10 + 10 + 10 = ☐

☐ tens

☐ times ☐

☐ × ☐ = ☐

☐ × ☐ = ☐

(d)

12 + 12 + 12 = ☐

☐ twelves

☐ times ☐

☐ × ☐ = ☐

☐ × ☐ = ☐

4 Draw lines to match the calculations with the number sentences.

 2 + 2 + 2 + 2

2 times 3

 3 + 3

4 times 11

 11 + 11 + 11 + 11

4 times 2

 6 + 6 + 6 + 6 + 6 + 6

6 times 6

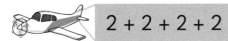

Challenge and extension question

5 Look at the shapes and then fill in each ◯ with a number.

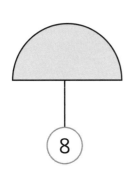

8

◯

◯

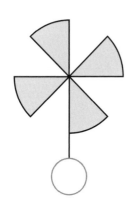

◯

◯

◯

5.7 Times (2)

 Learning objective Use grouping and repeated addition to model multiplication.

 Basic questions

1 Look at the pictures and write your answers in the spaces.

(a)

☐ groups of ☐ make ☐ .

Multiplication sentence:

_____

☐ times ☐ is ☐ .

(b)

☐ groups of ☐ make ☐ .

Multiplication sentence:

_____

☐ times ☐ is ☐ .

(c)

☐ groups of ☐ make ☐ .

Multiplication sentence:

_____

☐ times ☐ is ☐ .

(d)

☐ groups of ☐ make ☐ .

Multiplication sentence:

_____

☐ times ☐ is ☐ .

2 Count and then write the answers in the boxes.

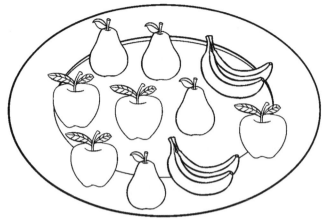

There are ☐ types of fruit. There are ☐ pieces in each type.

☐ groups of ☐ make ☐.

☐ times ☐ is ☐.

There are ☐ pieces of fruit in total.

3 First draw and then write the number sentences.

(a) △ △

The number of ◯ is 5 times the number of △.

Number sentence: _____

(b)

The number of ◯ is twice the number of ☆.

Number sentence: _____

(c)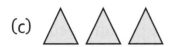

The number of ◯ is 3 times the number of △.

Number sentence: _____

4 Think carefully and then fill in the answers.

(a) ☺ ☺ ☺ ☺

The number of ♡ is 3 times the number of ☺.

The number of ♡ is ▢.

(b)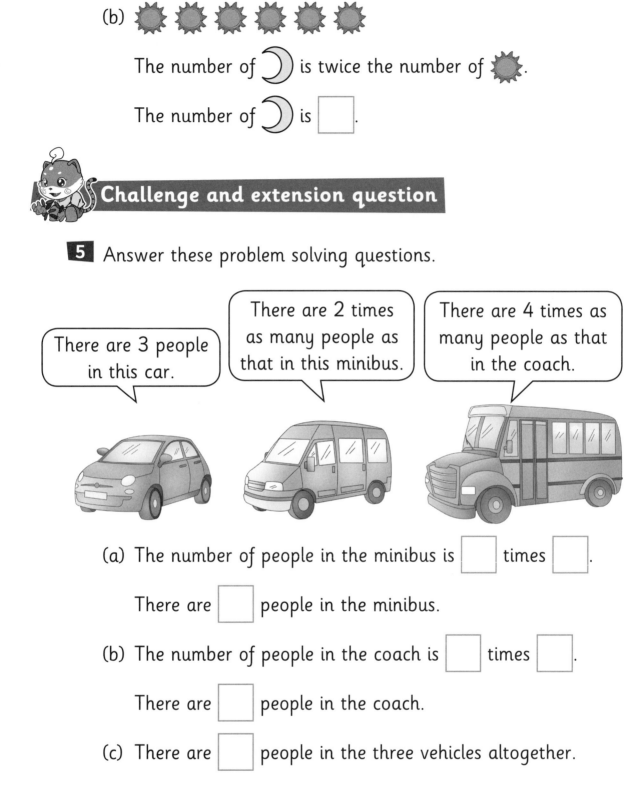

The number of ☽ is twice the number of ☀.

The number of ☽ is ☐.

Challenge and extension question

5 Answer these problem solving questions.

There are 3 people in this car.

There are 2 times as many people as that in this minibus.

There are 4 times as many people as that in the coach.

(a) The number of people in the minibus is ☐ times ☐.

There are ☐ people in the minibus.

(b) The number of people in the coach is ☐ times ☐.

There are ☐ people in the coach.

(c) There are ☐ people in the three vehicles altogether.

5.8 Multiplication of 10

Learning objective Recall and use the multiplication facts for the 10 times table

Basic questions

1 Look at the picture and fill in the boxes.

There are ☐ ☆ in a row.

There are ☐ rows.

We can say that it is ☐ groups of ☐ ,

or it is ☐ times ☐ .

Number sentence: ☐ × ☐ = ☐ .

There are ☐ stars in total.

2 Who can do it quickly and correctly?

9 × 10 = ☐   2 × 10 = ☐   4 × 10 = ☐   10 × 7 = ☐

10 × 1 = ☐   10 × 6 = ☐   10 × 8 = ☐   10 × 10 = ☐

5 × 10 = ☐   10 × 10 = ☐   0 × 10 = ☐   3 × 10 = ☐

3 Follow the number pattern on the first pizza slice to complete the rest.

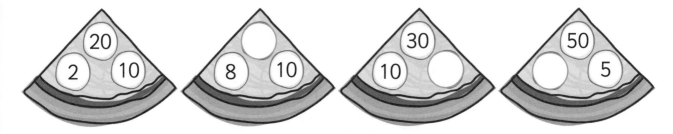

4 (a) Write a multiplication sentence to show how many apples there are.

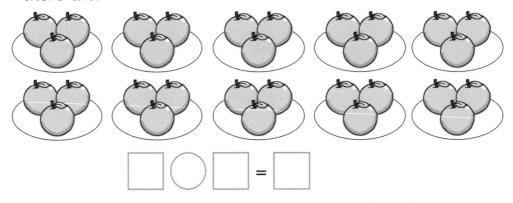

☐ ◯ ☐ = ☐

(b) Now how many apples are there on the plates?

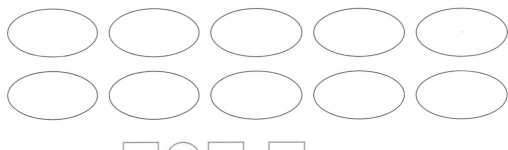

☐ ◯ ☐ = ☐

(c) Each  costs £10.

How much do 8 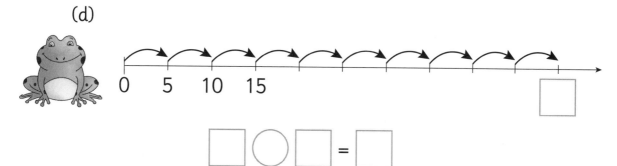 cost?

□ ○ □ = £ □

(d)

0   5   10   15   □

□ ○ □ = □

5 Think carefully and then fill in the answers.

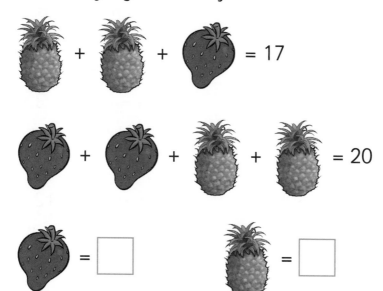

🍍 + 🍍 + 🍓 = 17

🍓 + 🍓 + 🍍 + 🍍 = 20

🍓 = □   🍍 = □

29

5.9 Multiplication of 5

Learning objective Recall and use the multiplication facts for the 5 times table

Basic questions

1 Write the numbers to complete the multiplication facts.

One times five is ☐.　　　　Two times five is ☐.

Three times five is ☐.　　　Four times five is ☐.

Five times five is ☐.　　　　Five times six is ☐.

Seven times five is ☐.　　　Eight times five is ☐.

Nine times five is ☐.　　　　Ten times five is ☐.

Eleven times five is ☐.　　　Twelve times five is ☐.

2 Calculate the answers to these questions.

5 × 4 = ☐　　5 × 6 = ☐　　8 × 5 = ☐　　3 × 5 = ☐

10 × 5 = ☐　　5 × 9 = ☐　　2 × 5 = ☐　　7 × 5 = ☐

5 × 8 = ☐　　9 × 5 = ☐　　5 × 11 = ☐　　12 × 5 = ☐

6 × 5 = ☐　　11 × 5 = ☐　　5 × 12 = ☐

3 Write a multiplication sentence for each picture.

(a) How many paintbrushes are there altogether?

$$\square \bigcirc \square = \square$$

(b) How many cupcakes are there altogether?

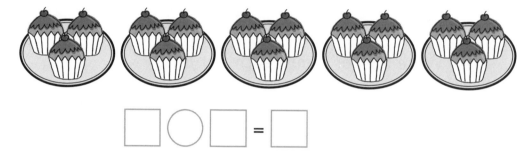

$$\square \bigcirc \square = \square$$

4 Write a multiplication sentence for each of the following.

(a) 1 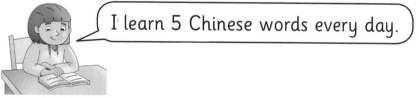 ate 5 🍌. How many 🍌 did 6 monkeys eat?

$$\square \bigcirc \square = \square$$

(b) How many Chinese words can Lily learn every week?

I learn 5 Chinese words every day.

$$\square \bigcirc \square = \square$$

(c) Each ⬭ can hold 8 🍎.

How many 🍎 can 5 ⬭ hold?

☐ ◯ ☐ = ☐

5 Complete the number pattern.

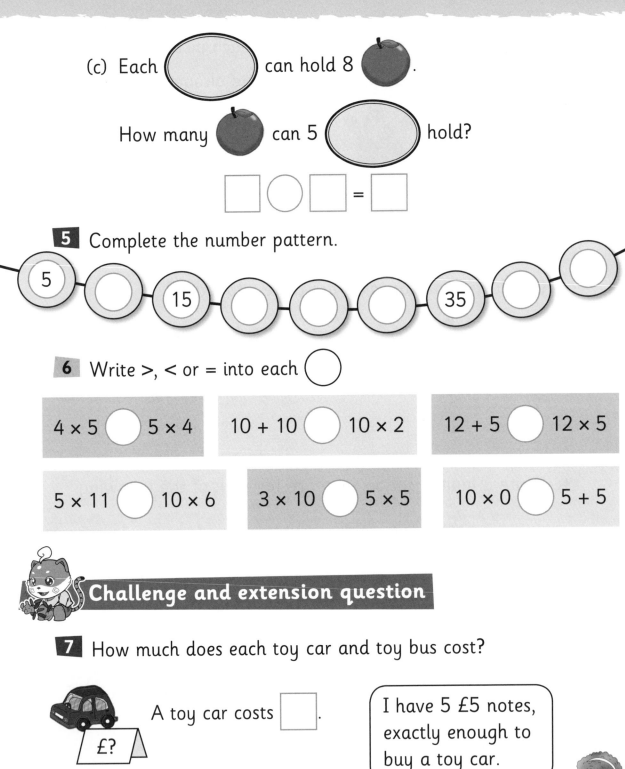

5 ◯ 15 ◯ ◯ ◯ 35 ◯ ◯

6 Write >, < or = into each ◯

| 4 × 5 ◯ 5 × 4 | 10 + 10 ◯ 10 × 2 | 12 + 5 ◯ 12 × 5 |

| 5 × 11 ◯ 10 × 6 | 3 × 10 ◯ 5 × 5 | 10 × 0 ◯ 5 + 5 |

Challenge and extension question

7 How much does each toy car and toy bus cost?

£? A toy car costs ☐.

I have 5 £5 notes, exactly enough to buy a toy car.

£? A toy bus costs ☐.

I need 1 more £5 note to buy a toy bus.

5.10 Multiplication of 2

Learning objective Recall and use the multiplication facts for the 2 times table

Basic questions

1 Complete the multiplication facts.

Two times one is ☐.   Two times two is ☐.

Two times three is ☐.   Two times four is ☐.

Two times five is ☐.   Two times six is ☐.

Two times seven is ☐.   Two times eight is ☐.

Two times nine is ☐.   Two times ten is ☐.

Two times eleven is ☐.   Two times twelve is ☐.

2 Who can do it quickly and correctly?

$2 \times 0 = $ ☐ $2 \times 6 = $ ☐ $8 \times 5 = $ ☐ $3 \times 2 = $ ☐

$5 \times 2 = $ ☐ $2 \times 2 = $ ☐ $7 \times 2 = $ ☐ $2 \times 3 = $ ☐

$2 \times 8 = $ ☐ $6 \times 5 = $ ☐ $2 \times 1 = $ ☐ $5 \times 10 = $ ☐

$2 \times 12 = $ ☐ $11 \times 5 = $ ☐ $2 \times 11 = $ ☐ $5 \times 12 = $ ☐

3 Look at the pictures and write the answers.

(a) Counting the rows, there are ☐ groups of ☐.

Multiplication sentence: ☐ × ☐ = ☐

(b) Counting the columns, there are ☐ groups of ☐.

Multiplication sentence: ☐ × ☐ = ☐

4 Write a multiplication sentence for each picture.

(a) How many ladybirds are there altogether?

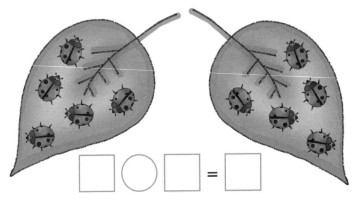

☐ ◯ ☐ = ☐

(b) How many pears are there altogether?

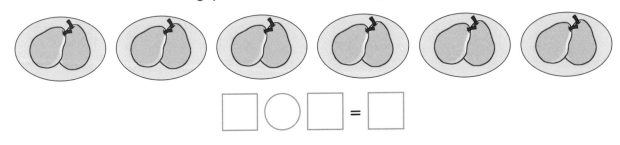

☐ ◯ ☐ = ☐

(c) How many shoes are there altogether?

$\boxed{}\bigcirc\boxed{} = \boxed{}$

(d) One  costs £5. The cost of a is 2 times as

much as a  .

How much does a cost?

$\boxed{}\bigcirc\boxed{} = £\boxed{}$

5 How much will it cost to go to the fair?

Ticket Price

Adult: £5 per person

Child: £2 per person

(a) To buy 8 children's tickets:

$\boxed{}\bigcirc\boxed{} = \boxed{}$ pounds → £$\boxed{}$

(b) To buy 12 adult tickets:

$\boxed{}\bigcirc\boxed{} = \boxed{}$ pounds → £$\boxed{}$

(c) To buy 10 children's tickets and 10 adult tickets:

$\boxed{}\bigcirc\boxed{} = \boxed{}$ pounds → £$\boxed{}$

Challenge and extension questions

6 Calculate. Note that when a number sentence includes both multiplication and addition or subtraction, we do multiplication before addition or subtraction.

$8 + 5 \times 2 =$ ☐ $7 \times 2 + 16 =$ ☐

$2 \times 10 - 6 =$ ☐ $5 + 11 \times 2 =$ ☐

$28 - 8 \times 2 =$ ☐ $52 - 0 \times 2 =$ ☐

7 Think carefully. How many umbrellas are there?

Do you have a better way than counting to find out?

Number sentence: _____

Number sentence: _____

5.11 Multiplication of 4

 Learning objective Recall and use the multiplication facts for the 4 times table

 Basic questions

1 Calculate mentally.

8 × 4 = ☐   3 × 4 = ☐   2 × 9 = ☐   4 × 7 = ☐

0 × 4 = ☐   5 × 4 = ☐   10 × 4 = ☐   5 × 6 = ☐

4 × 1 = ☐   9 × 4 = ☐   2 × 5 = ☐   4 × 4 = ☐

4 × 11 = ☐   12 × 4 = ☐   10 × 5 = ☐   4 × 12 = ☐

2 Fill in the boxes with suitable numbers.

☐ × 4 = 28   4 × ☐ = 4   4 × ☐ = 0

4 × ☐ = 40   ☐ × 8 = 40   ☐ × 9 = 18

3 × ☐ = 12   4 × ☐ = 36   4 × ☐ = 16

4 × ☐ = 48   4 × ☐ = 20   ☐ × 4 = 44

3 Write a +, – or × into each ◯ to make the number sentence true.

4 ◯ 4 = 16   4 ◯ 4 = 0

4 ◯ 4 = 8   4 ◯ 0 = 4

4 Draw a line to match each alien with a spaceship.

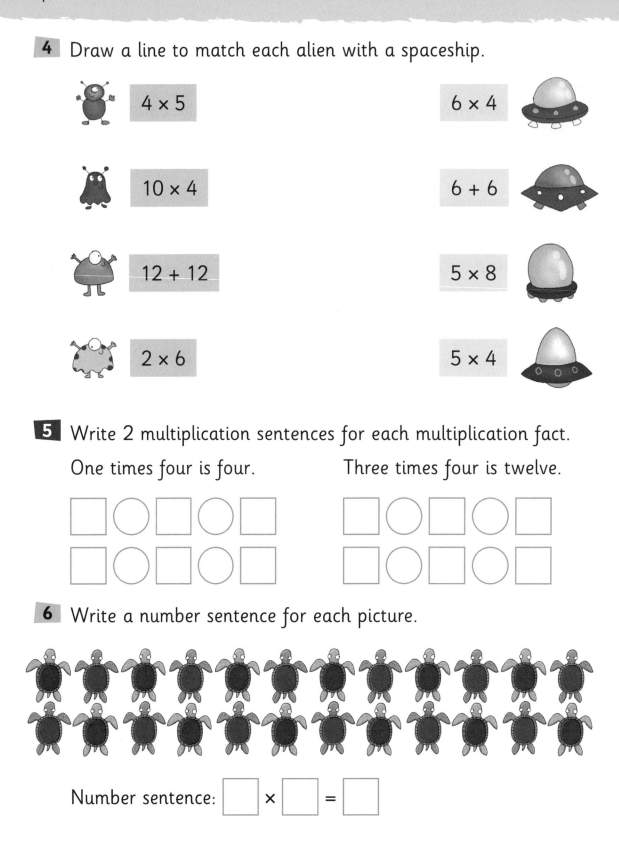

4 × 5   6 × 4

10 × 4   6 + 6

12 + 12   5 × 8

2 × 6   5 × 4

5 Write 2 multiplication sentences for each multiplication fact.

One times four is four.   Three times four is twelve.

6 Write a number sentence for each picture.

Number sentence: ☐ × ☐ = ☐

[ ] wheels

Number sentence: [ ] × [ ] = [ ]

7 Answer these questions.

(a) There are 10 in a grass field.

There are 4 times as many  as  .

How many  are there?

Number sentence: [ ] × [ ] = [ ]

(b) Each costs £8.

How much do 4 cost?

Number sentence: [ ] × [ ] = £[ ]

Challenge and extension question

8 What mathematical questions can you make up for each picture?

Write a number sentence to show your answer.

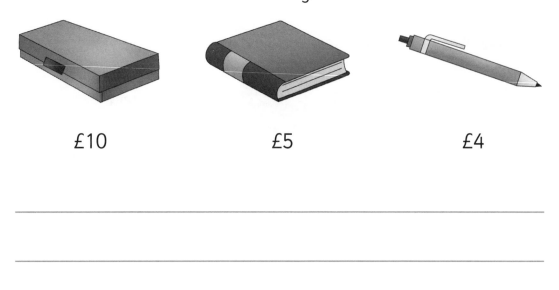

£10   £5   £4

_____

_____

_____

5.12 Multiplication of 8

 Learning objective Recall and use the multiplication facts for the 8 times table

 Basic questions

1 Complete the multiplication facts.

Two times _____ is sixteen.

Four times eight is _____.

_____ times _____ is twenty-four.

Five times _____ is forty.

Six times eight is _____.

_____ times _____ is fifty-six.

_____ times eleven is eighty-eight.

Eight times twelve is _____.

2 Fill in the boxes with suitable numbers.

$5 \times \boxed{} = 40$ $\boxed{} \times 6 = 48$ $\boxed{} \times 8 = 16$

$7 \times \boxed{} = 56$ $\boxed{} \times 8 = 80$ $8 \times \boxed{} = 24$

$8 \times \boxed{} = 72$ $\boxed{} \times 8 = 16$ $\boxed{} \times 8 = 64$

$8 \times \boxed{} = 96$ $8 \times \boxed{} = 8$ $\boxed{} \times 8 = 88$

3 Write >, < or = into each ◯

$4 + 8$ ◯ 12 3×8 ◯ 8×3 8×6 ◯ 50

8×8 ◯ 16 2×8 ◯ $2 + 8$ 0×8 ◯ 8

4 Write a multiplication sentence for each picture.

(a)

☐ ◯ ☐ = ☐

(b)

☐ ◯ ☐ = ☐

(c) I eat 8 carrots every day. How many carrots can I eat in a week?

☐ ◯ ☐ = ☐

5 (a) One cat has 4 legs. How many legs do 8 cats have altogether?

Number sentence: _____

(b) People are rowing on the river.
There are 6 people in each boat.
How many people are there in 8 boats?

Number sentence: _____

(c) Emma is 8 years old.
Her gran is 8 times as old as Emma.
How old is Emma's gran?

Number sentence: _____

6 Calculate the answers to these questions.

$5 \times 8 - 5 =$ ☐ $8 \times 8 + 8 =$ ☐ $8 \times 0 + 8 =$ ☐

$10 \times 8 + 2 =$ ☐ $9 \times 8 + 9 =$ ☐ $1 \times 8 + 8 =$ ☐

$2 \times 4 + 8 =$ ☐ $5 \times 5 + 5 =$ ☐ $10 \times 0 + 0 =$ ☐

$11 \times 5 + 8 =$ ☐ $5 \times 12 + 5 =$ ☐ $8 \times 11 + 12 =$ ☐

Challenge and extension question

7 Serena walked from the first tree to the ninth tree.

How many metres did she walk?

There are 5 metres between
every two neighbouring trees.

Serena walked ☐ metres.

5.13 Relationships between multiplications of 2, 4 and 8

Learning objective Use the relationship between the 2, 4 and 8 times tables

 **Basic questions**

1 Fill in the table.

×	1	2	3	4	5	6	7	8	9	10	11	12
2												
4												
8												

1 eight = ☐ fours = ☐ twos

2 eights = ☐ fours = ☐ twos

5 eights = ☐ fours = ☐ twos

11 eights = ☐ fours = ☐ twos

2 Complete each multiplication fact.

☐ × 2 = 4	2 × ☐ = 8	16 = ☐ × 2	20 = ☐ × 2	24 = ☐ × 1
☐ × 4 = 4	4 × ☐ = 8	16 = ☐ × 4	20 = ☐ × 4	24 = ☐ × 3
	8 × ☐ = 8	16 = ☐ × 4	20 = ☐ × 5	24 = ☐ × 8
		16 = ☐ × 8	20 = ☐ × 10	24 = ☐ × 4
			20 = ☐ × 20	24 = ☐ × 6
				24 = ☐ × 24

3 Calculate first and then find patterns.

2 × 2 = ☐   2 × 4 = ☐   6 × 2 = ☐

4 × 2 = ☐   4 × 4 = ☐   6 × 4 = ☐

8 × 2 = ☐   8 × 4 = ☐   6 × 8 = ☐

8 × 2 = ☐   10 × 2 = ☐   12 × 2 = ☐

8 × 4 = ☐   10 × 4 = ☐   12 × 4 = ☐

8 × 8 = ☐   10 × 8 = ☐   12 × 8 = ☐

What patterns did you find?

4 (a) 4 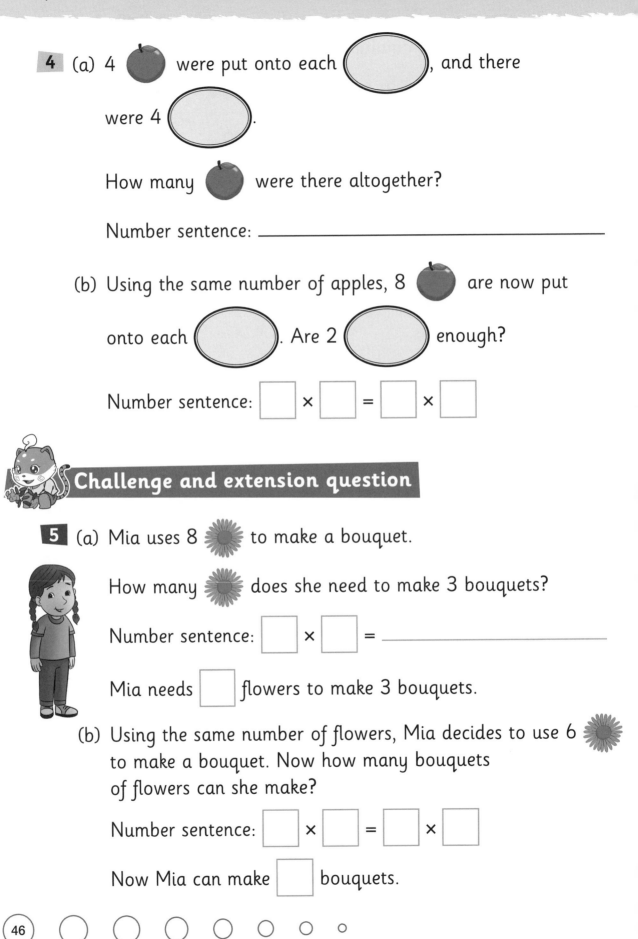 were put onto each ( ), and there

were 4 ( ).

How many were there altogether?

Number sentence: _____

(b) Using the same number of apples, 8 are now put

onto each ( ). Are 2 ( ) enough?

Number sentence: ☐ × ☐ = ☐ × ☐

Challenge and extension question

5 (a) Mia uses 8 to make a bouquet.

How many does she need to make 3 bouquets?

Number sentence: ☐ × ☐ = _____

Mia needs ☐ flowers to make 3 bouquets.

(b) Using the same number of flowers, Mia decides to use 6
to make a bouquet. Now how many bouquets
of flowers can she make?

Number sentence: ☐ × ☐ = ☐ × ☐

Now Mia can make ☐ bouquets.

5.14 Dividing and division

 Learning objective Use grouping and repeated subtraction to divide

Basic questions

1 Make a drawing first. Then fill in each box.

(a) A box holds 4 buns. How many boxes are needed to hold 12 buns?

$12 - \boxed{} - \boxed{} - \boxed{} = 0$

$12 = \boxed{} \times 4$

$12 \div 4 = \boxed{}$

(b) How many can be made with ● in the diagram below?

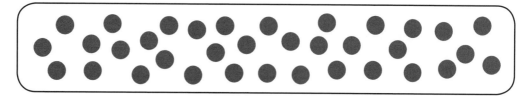

$\boxed{} - \boxed{} - \boxed{} - \boxed{} - \boxed{} - \boxed{} - \boxed{} - \boxed{} = 0$

$\boxed{} = \boxed{} \times \boxed{}$

$\boxed{} \div \boxed{} = \boxed{}$

2 Look at the picture and fill in the ☐.

(a) There are ☐ sweets altogether.

(b) If each child gets 2 sweets, ☐ children will have sweets.

(c) If each child gets 3 sweets, ☐ children will have sweets.

(d) If each child gets 4 sweets, ☐ children will have sweets

and ☐ sweets will be left over.

3 Amir uses a 4-metre-long tape to measure the rope. How many times does he need to measure to find out the length of the whole rope?

16 m

4 m

Division sentence: ☐ ÷ ☐ = ☐

4 ☆ ☆ ☆ ☆ ☆ ☆ ☆ ☆

(a) Divide the stars into 2 equal groups. How many stars does each group have?

☐ ÷ ☐ = ☐

(b) Divide the stars into groups of 4.
How many groups can they be divided into?

☐ ÷ ☐ = ☐

5

(a) Divide the mangoes into 3 equal groups. How many mangoes are there in each group?

$\boxed{} \div \boxed{} = \boxed{}$

(b) If you put 5 mangoes on a plate, how many plates are needed?

$\boxed{} \div \boxed{} = \boxed{}$

Challenge and extension question

6

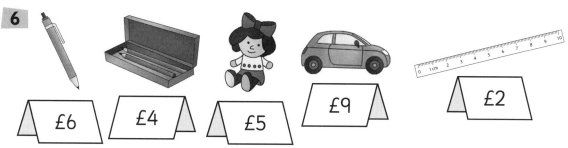

£6   £4   £5   £9   £2

(a) How many pencil cases can Shafiq buy with £20?

_____

Shafiq can buy $\boxed{}$ pencil cases.

(b) How many dolls can Minna buy with £30?

_____

Minna can buy $\boxed{}$ dolls.

(c) Write two more questions using the items on the previous page.

Show the answer for each one.

Question 1:

_____

_____

Answer:

_____

Question 2:

_____

_____

Answer:

_____

5.15 Finding quotients using multiplication facts

Learning objective Use multiplication facts to find quotients

Basic questions

1 Use multiplication facts to find quotients.

(a) Two times ☐ is sixteen.

16 ÷ 2 = ☐

(b) Five times ☐ is thirty.

30 ÷ 5 = ☐

(c) ☐ times eight is thirty-two.

32 ÷ 8 = ☐

(d) Four times ☐ is thirty-six.

36 ÷ 4 = ☐

2 Use each multiplication fact to write two multiplication sentences and two division sentences.

Three times seven is twenty-one.

_____

_____

_____

_____

Five times six is thirty.   Seven times eight is fifty-six.

_____   _____

_____   _____

_____   _____

3 Draw a line to make each duck reach the pond.

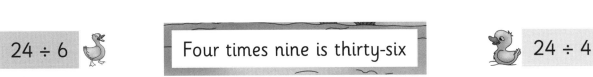

35 ÷ 7   Four times six is twenty-four   36 ÷ 4

24 ÷ 6   Four times nine is thirty-six   24 ÷ 4

24 ÷ 8   Five times seven is thirty-five   35 ÷ 5

36 ÷ 9   Three times eight is twenty-four   24 ÷ 3

4 Who can do it quickly and correctly?

12 ÷ 4 = ☐   20 ÷ 5 = ☐   40 ÷ 5 = ☐   20 ÷ 2 = ☐

32 ÷ 8 = ☐   16 ÷ 8 = ☐   36 ÷ 9 = ☐   24 ÷ 4 = ☐

10 ÷ 2 = ☐   28 ÷ 4 = ☐   64 ÷ 8 = ☐   72 ÷ 8 = ☐

45 ÷ 5 = ☐   18 ÷ 2 = ☐   40 ÷ 10 = ☐   30 ÷ 5 = ☐

44 ÷ 11 = ☐   96 ÷ 8 = ☐   36 ÷ 12 = ☐   77 ÷ 7 = ☐

5 Write a suitable number into each ☐.

$$\boxed{} \div 2 = 5 \qquad \boxed{} \div 5 = 6 \qquad \boxed{} \div 8 = 1 \qquad \boxed{} \div 4 = 0$$

$$\boxed{} \div 4 = 4 \qquad \boxed{} \div 4 = 7 \qquad \boxed{} \div 5 = 7 \qquad \boxed{} \div 10 = 4$$

$$\boxed{} \div 5 = 3 \qquad \boxed{} \div 2 = 10 \qquad \boxed{} \div 4 = 9 \qquad \boxed{} \div 8 = 8$$

6 Fill in the ◯ with >, < or =.

$48 \div 8 \bigcirc 12$	$30 \div 5 \bigcirc 12 \div 2$	$20 \div 2 \bigcirc 2 \times 5$
$25 \div 5 \bigcirc 6$	$24 \div 4 \bigcirc 24 \div 8$	$4 \times 3 \bigcirc 32 \div 8$
$18 \div 2 \bigcirc 10$	$35 \div 5 \bigcirc 40 \div 5$	$28 \div 4 \bigcirc 2 + 5$
$55 \div 11 \bigcirc 6$	$84 \div 7 \bigcirc 66 \div 6$	$96 \div 12 \bigcirc 2 \times 4$

Challenge and extension question

7 Work out the missing number.

🕷 × 🕷 + 🕷 = 6

🐌 × 🐌 + 🐌 = 72

🐌 ÷ 🕷 = ☐

5.16 How many times?

 Learning objective Multiply by scaling up, using the language of 'times'

 Basic questions

1 Look at each picture and answer the questions.

(a)

The number of ☆ is ⬜ times the number of ☽.

Number sentence: _____

(b) The first row

The second row

The number of 🌞 in the second row is ⬜ times the number of 🌞 in the first row.

Number sentence: _____

(c)  8   24

The number of bees is ⬜ times the number of butterflies.

Number sentence: _____

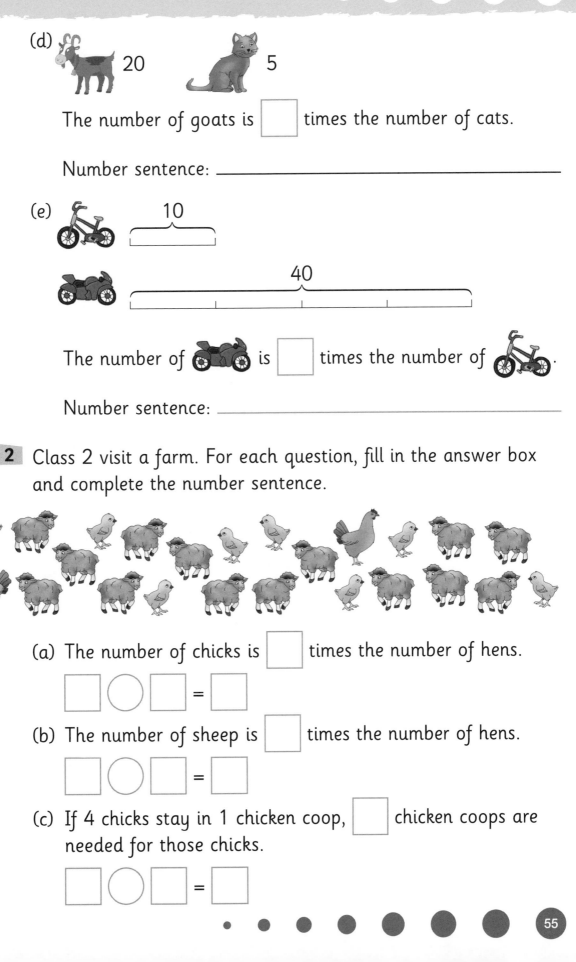

(d) 20   5

The number of goats is ⬚ times the number of cats.

Number sentence: _____

(e) 10

40

The number of 🏍 is ⬚ times the number of 🚲.

Number sentence: _____

2 Class 2 visit a farm. For each question, fill in the answer box and complete the number sentence.

(a) The number of chicks is ⬚ times the number of hens.

⬚ ◯ ⬚ = ⬚

(b) The number of sheep is ⬚ times the number of hens.

⬚ ◯ ⬚ = ⬚

(c) If 4 chicks stay in 1 chicken coop, ⬚ chicken coops are needed for those chicks.

⬚ ◯ ⬚ = ⬚

3 (a) There are 40 volleyballs and 5 footballs in the sports room. How many times as many footballs are there as volleyballs?

Answer: _____

(b) Aaron is 8 years old. His mother is 32 years old. How many times as old as Aaron is his mother?

Answer: _____

Challenge and extension question

4

(a) If 1 more grey rabbit and 1 one more white rabbit join them, will the number of grey rabbits still be 2 times the number of white rabbits?

Answer: _____

(b) If 1 more white rabbit joins, how many grey rabbits should also join so that the number of grey rabbits is still 2 times the number of white rabbits?

Answer: _____

5.17 Division with dividend 0

 **Learning objective** Multiply by zero and divide with a dividend of zero

Basic questions

1 Write a division sentence for each picture.

(a) Share 20 turtles equally between 4 ponds. How many turtles are there in each pond?

Division sentence: _____

(b) When there are no turtles, how many turtles can be shared by each pond?

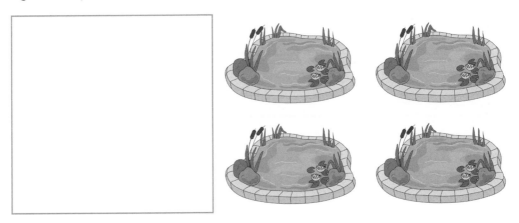

Division sentence: _____

2 Who can do it quickly and correctly?

$0 \times 8 = \boxed{}$　　$8 - 0 = \boxed{}$　　$0 + 21 = \boxed{}$　　$40 \times 1 = \boxed{}$

$5 + 0 = \boxed{}$　　$10 \times 0 = \boxed{}$　　$29 - 29 = \boxed{}$　　$0 + 100 = \boxed{}$

$0 \div 2 = \boxed{}$　　$0 \div 8 = \boxed{}$　　$0 \div 1 = \boxed{}$　　$45 \div 1 = \boxed{}$

$0 \div 5 = \boxed{}$　　$0 \div 10 = \boxed{}$　　$0 \div 9 = \boxed{}$　　$0 \div 100 = \boxed{}$

$0 \div 11 = \boxed{}$　　$12 \times 0 = \boxed{}$　　$11 \div 11 = \boxed{}$　　$0 \div 12 = \boxed{}$

3 Complete this table.

Dividend	0	35	0	28		64		60	
Divisor	4	5	10		16		12		11
Quotient				4	0	8	0	5	0

4 Write a +, −, × or ÷ into each ◯ to make the number sentence true.

5 ◯ 6 = 11	20 ◯ 20 = 1	5 ◯ 0 = 5	12 ◯ 3 = 4
0 ◯ 10 = 0	0 ◯ 8 = 0	8 ◯ 8 = 64	8 ◯ 10 = 80
35 ◯ 5 = 30	24 ◯ 8 = 32	24 ◯ 0 = 0	0 ◯ 100 = 0
33 ◯ 3 = 11	99 ◯ 11 = 88	0 ◯ 7 = 0	7 ◯ 11 = 77

5 Write a number sentence for each question.

(a) The dividend is 0 and the divisor is 4. What is the quotient?

Number sentence: _____

(b) What is the product when 2 eights are multiplied?

Number sentence: _____

(c) How many times greater is 45 than 9?

Number sentence: _____

(d) How many times 5 is 25?

Number sentence: _____

(e) 72 marbles are divided into 8 groups equally. How many marbles are there in each group?

Number sentence: _____

(f) One factor is 6. The other factor is 10. What is the product?

Number sentence: _____

Challenge and extension question

6 What is the greatest number you can write in each box?

3 × ☐ < 18	☐ × 5 < 41	☐ × 8 < 42	48 − ☐ > 40
5 × ☐ < 32	☐ × 4 < 28	2 × ☐ < 20	☐ × 5 < 3 × 8
11 × ☐ < 100	☐ × 4 < 3	12 × ☐ < 59	☐ × 6 < 8 × 9

5.18 Practice and exercise (III)

Learning objective Recall and use the multiplication facts for 2, 4, 5, 8 and 10 to solve problems

Basic questions

1 Look at each picture and then fill in the answers.

Number of ducks:	Number of ladybirds:
☐ groups of ☐ ,	☐ groups of ☐ ,
or ☐ times ☐ .	or ☐ times ☐ .
Addition sentence:	Addition sentence:
_____	_____
Multiplication sentence:	Multiplication sentence:
_____	_____

How many more ladybirds are there than ducks?

Number sentence: _____

There are ☐ more ladybirds than ducks.

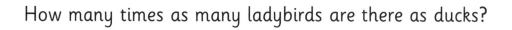

How many times as many ladybirds are there as ducks?

Number sentence: _____

There are ☐ times as many ladybirds as ducks.

2 Link each addition sentence with a multiplication sentence.

| 4 + 4 + 4 | 5 + 5 | 2 + 2 + 2 + 2 |

| 2 × 5 | 4 × 2 | 4 × 3 |

3 Draw the beads onto the string. The first has been done for you.

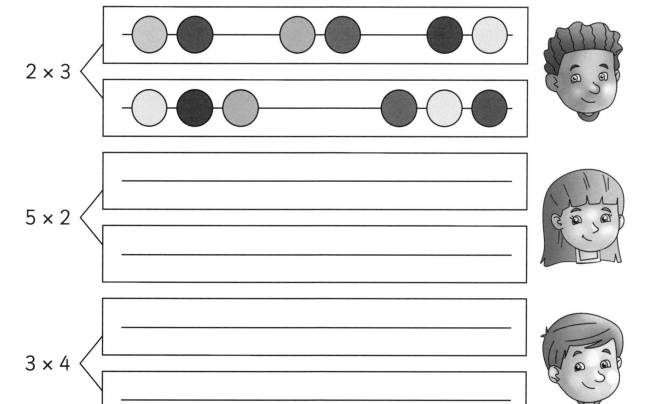

2 × 3

5 × 2

3 × 4

4 Look at the picture. How many multiplication sentences can you write?

_____

_____

5 Think carefully and then write the number sentence.

(a) If 2 cakes are packed in a box, how many cakes are there in 10 boxes?

Number sentence: _____

(b) If one box can contain exactly 2 cakes, how many boxes are needed to contain 10 cakes?

Number sentence: _____

(c) There are 8 chocolate cakes. There are 3 times as many cherry cakes as chocolate cakes. How many cherry cakes are there?

Number sentence: _____

6 Fill in the table.

×	2	4	6	8	10	12	11	9	7	5	3
2											
4											
5											
8											
10											

Challenge and extension question

7 This is a price list:

Book: £7

Pencil case: £3

Art kit: £18

Computer game: £30

Use this information to make up your own questions.

(a) Addition question: _____

_____

Number sentence: _____

(b) Subtraction question: _____

_____

Number sentence: _____

(c) Division question: _____

_____

Number sentence: _____

Chapter 5 test

1 Calculate mentally.

21 − 7 = ☐   40 + 26 = ☐   25 + 20 = ☐   88 − 80 = ☐

18 ÷ 2 = ☐   10 ÷ 2 = ☐   5 × 9 = ☐   56 + 12 = ☐

36 − 13 = ☐   98 − 18 = ☐   20 ÷ 4 = ☐   9 × 1 = ☐

2 × 5 = ☐   90 − 10 = ☐   58 − 34 = ☐   7 × 5 = ☐

25 ÷ 5 = ☐   10 × 5 = ☐   69 + 1 = ☐   36 ÷ 9 = ☐

2 × 12 = ☐   11 × 7 = ☐   22 + 11 = ☐   84 ÷ 12 = ☐

2 Look at the pictures and then fill the answers in the boxes.

(a)

There are ☐ 🐟 in each row.

There are ☐ rows.

There are ☐ groups of ☐ .

We can also say it is ☐ times ☐ .

There are ☐ 🐟 in each column.

There are ☐ columns.

There are ☐ groups of ☐ .

We can also say it is ☐ times ☐ .

(b)

There are ☐  in each set.

There are ☐ sets.

There are ☐ 🐸 altogether.

Number sentence: ☐ × ☐ = ☐

(c)

There are ☐ 🦘 in the first 2 sets.

There are ☐ 🦘 in the last set.

Number sentence: ☐ × ☐ + ☐ = ☐

3 Work out the answers to these calculations.

5 × 8 = ☐   4 + 4 + 4 + 4 + 4 = ☐

3 × 10 = ☐   2 + 2 + 2 + 2 + 2 + 2 + 2 + 2 + 2 = ☐

11 × 6 = ☐   12 + 12 + 12 + 12 = ☐

4 times 9 = _____

7 times 8 = _____

4 Follow the information to draw the correct number of each shape.

(a) There are 10 ☆. The number of △ is half of that of ☆. Draw ☆ and △.

_____

_____

(b) There are 5 ♡. The number of ♡ is 2 fewer than that of ◇. Draw ♡ and ◇.

_____

_____

(c) There are 4 ◇. There are 3 times as many ◯ as ◇. Draw ◯ and ◇.

_____

_____

(d) There are 3 ☆ in each set. There are 3 sets of ☆. Draw ☆.

_____

5 Think carefully and then solve each word problem.

(a) There are 10 books on each shelf in a bookcase. There are 5 shelves. How many books are there in the bookcase?

Number sentence: _____

(b) Amaya folds paper stars. She folds 4 paper stars with 1 piece of paper. How many paper stars can she fold with 8 pieces of such paper?

Number sentence: _____

(c) There are 5 vans in the car park. There are 6 times as many cars as vans. How many cars are there in the car park?

Number sentence: _____

(d) How many tricycles can be made using 21 wheels?

Number sentence: _____

(e) 28 pieces of chocolate are equally shared by Lucas and his 3 friends. How many pieces of chocolate can each person get?

Number sentence: _____

6 Thirty pupils take part in art activities.

(a) Divide the pupils into 5 equal groups. How many pupils are there in each group?

Answer: _____

(b) The teacher in charge needs to prepare 2 paper bags for each pupil and 5 extra as a backup. How many paper bags does the teacher need to prepare for the art activities?

Answer: _____

(c) Each group gets 2 stacks of drawing paper. Each stack costs £10. How much is the cost of the drawing paper in total?

Answer: _____

Chapter 6 Introduction to fractions

6.1 Whole and part

 Learning objective Recognise fractions of a shape or quantity as a part of a whole

 Basic questions

1 Which side is the 'whole' and which side is a 'part' of the whole? Write 'whole' or 'part' underneath each one.

2 Divide the whole into parts and then draw one part in the box on the right.

(a)

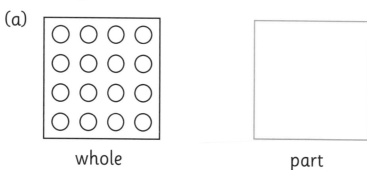

whole   part

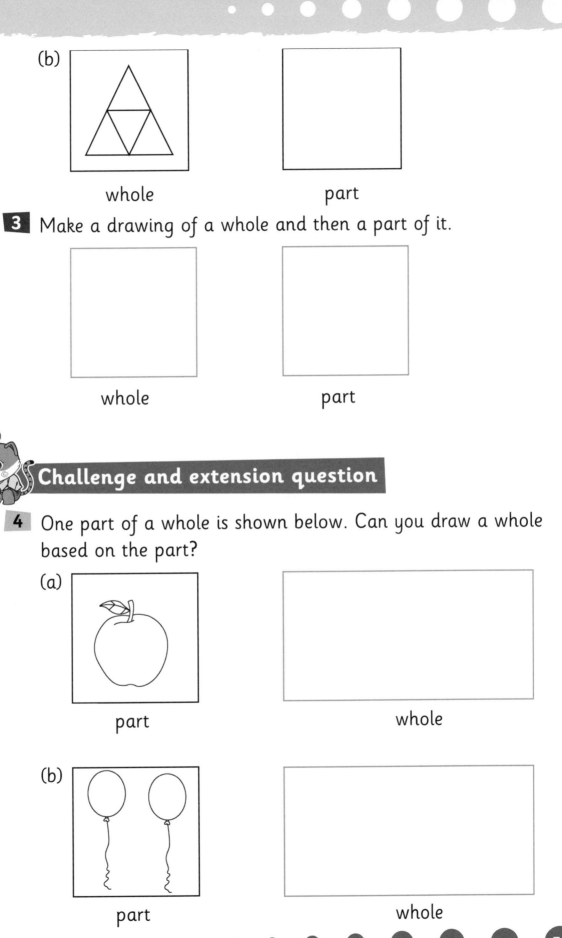

(b)

whole

part

3 Make a drawing of a whole and then a part of it.

whole

part

Challenge and extension question

4 One part of a whole is shown below. Can you draw a whole based on the part?

(a)

part

whole

(b)

part

whole

6.2 Unit fractions (1)

Learning objective Use fraction notation to show unit fractions

Basic questions

1 Have these objects been divided equally? Put a tick (✓) for yes and a cross (✗) for no.

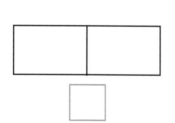 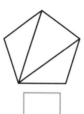  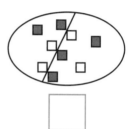

2 Fill in the blanks and write a fraction to show each shaded part. The first one has been done for you.

(a)

The rectangle is divided into __4__ parts.

The shaded part is ____ of the whole.

(b)

The rectangle is divided into ____ parts.

The shaded part is ____ of the whole.

(c) Write a fraction to show each shaded part of the whole.

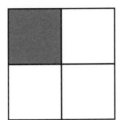   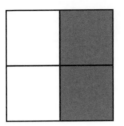

_____   _____

3 Colour or circle the part(s) according to the fractions given.

      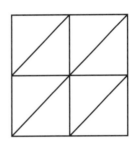

$\dfrac{1}{3}$ $\dfrac{3}{4}$ $\dfrac{1}{8}$

4 Express the shaded part of each diagram as a fraction.

      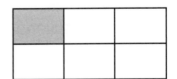

_____   _____   _____

Challenge and extension question

5 Collect the information and fill in the boxes.

There are ____ pupils in my maths class.

I am one of these ____ pupils, or in terms of a fraction,

I am ____ of the whole class.

6 I am reading a book. On the first day, I read half of the whole book.

On the second day, I read half of the rest of the book.

What fraction of the book did I read in

the two days? Answer: ____

6.3 Unit fractions (2)

Learning objective Compare unit fractions and find fractions of amounts

Basic questions

1 Fill in the spaces.

(a) Numbers like $\frac{1}{2}$, $\frac{1}{3}$, $\frac{3}{4}$ ... are called _____.

(b) From the same whole, the more parts it is divided into equally, the _____ each part becomes.

The fewer parts it is divided into equally, the

_____ each part becomes.

(c) Compare the following fractions. Fill in the circles with >, < or = .

$\frac{1}{3}$ ◯ $\frac{1}{4}$ $\frac{1}{8}$ ◯ $\frac{1}{2}$ $\frac{1}{7}$ ◯ 1 $\frac{1}{2}$ ◯ $\frac{2}{4}$

2 Fill in the boxes on the number lines.

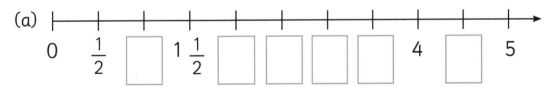

(a)

0    $\frac{1}{2}$ ☐ $1\frac{1}{2}$ ☐ ☐ ☐ ☐   4   ☐   5

(b)

1    $1\frac{1}{4}$ 2   ☐ ☐ $3\frac{1}{2}$ ☐ ☐ ☐

3 Write each of the following fractions using numbers. The first one has been done for you.

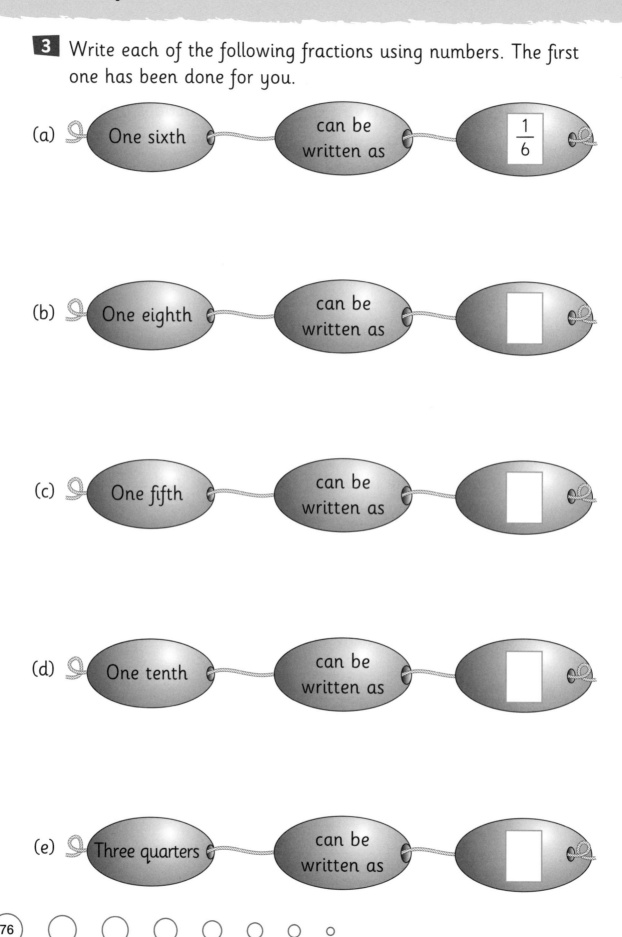

(a) One sixth — can be written as — $\dfrac{1}{6}$

(b) One eighth — can be written as — ☐

(c) One fifth — can be written as — ☐

(d) One tenth — can be written as — ☐

(e) Three quarters — can be written as — ☐

4 What fraction of the whole is the circled part in each diagram below?

Read it out and then fill in the answer line with words.

(a)

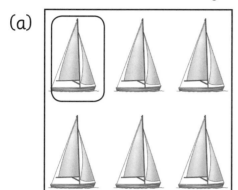

(b)

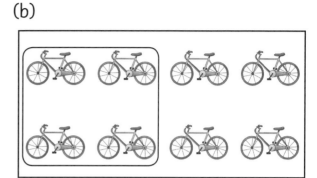

_____

It is read as

_____

_____

It is read as

_____

(c)

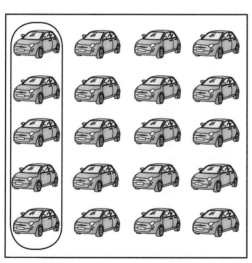

(d)

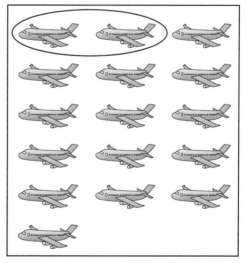

_____

It is read as

_____

_____

It is read as

_____

5 Colour the items in each picture below to show the given fraction.

$$\frac{1}{3}$$

$$\frac{1}{4}$$

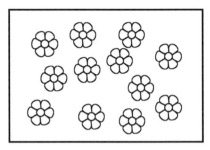

$$\frac{3}{4}$$

 Challenge and extension questions

6 Answer these multiple choice questions. Write your answer in the box.

(a) The picture below shows that the area of the shaded part is ☐ of the whole.

A. $\frac{1}{4}$ B. $\frac{1}{6}$ C. $\frac{1}{8}$ D. Can't be expressed using a fraction

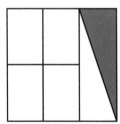

(b) To fold a rectangular piece of paper consecutively in equal halves to obtain a $\frac{1}{16}$ part of it, it should be folded ☐ times.

A. 2   B. 4   C. 8   D. 16

7 Answer these questions.

(a) $\frac{1}{2}$ of £6 is £ [ ] .

(b) $\frac{3}{4}$ of 12 apples is [ ] apples.

8 If a 10-metre-long rope is cut into 10 equal pieces, each piece

is [ ] of 10 metres.

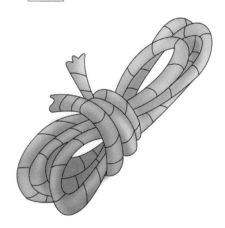

Each piece is [ ] metre long.

Chapter 6 test

1 Circle the items in each diagram to show the given fraction.

(a)

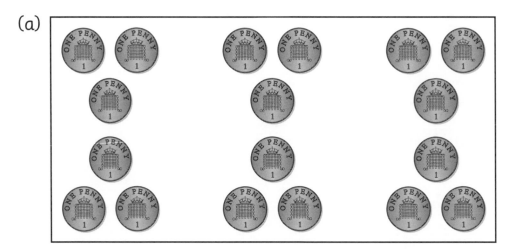

$$\frac{1}{3}$$

(b)

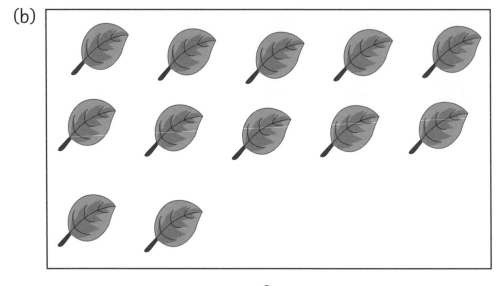

$$\frac{3}{4}$$

2 Count in fractions to complete the number lines.

(a) Count in fractions of $\frac{1}{2}$ and fill in the boxes.

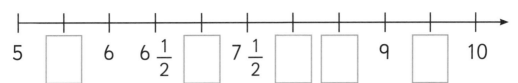

5 ☐ 6 $6\frac{1}{2}$ ☐ $7\frac{1}{2}$ ☐ ☐   9   ☐   10

(b) Count in fractions of $\frac{1}{4}$ and fill in the boxes.

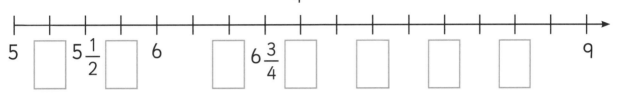

5 ☐ $5\frac{1}{2}$ ☐ 6   ☐ $6\frac{3}{4}$ ☐   ☐   ☐   ☐   9

3 Write the answers to these problems in the boxes.

(a) If a 10-metre-long rope is cut into ten equal pieces, each
piece is ☐ of 10 metres.

The length of each piece is ☐ metre.

If a 9-metre-long rope is cut into three equal pieces, each
piece is ☐ of 9 metres. Each piece is ☐ metres long.

(b) $\frac{1}{5}$ of the 10 △ is ☐ △.

△ △ △ △ △ △ △ △ △ △

$\frac{1}{3}$ of the ☐ ☆ is 4 ☆.

☆ ☆ ☆ ☆ ☆ ☆ ☆ ☆ ☆ ☆ ☆ ☆

(c) Compare the following fractions. Write >, < or = in
each circle.

| $\frac{1}{2}$ ◯ $\frac{1}{3}$ | $\frac{3}{4}$ ◯ 1 | $\frac{1}{4}$ ◯ $\frac{1}{2}$ |

4 True or false? Put a tick (✓) for true or a cross (✗) for false.

(a) When a piece of paper is folded in half for the first time, and then in half again for two more times, each part of the finally folded paper is $\frac{1}{3}$ of its original paper. ☐

(b) In the pictures below, both of the unshaded parts are $\frac{1}{6}$, therefore both the unshaded parts are of the same size. ☐

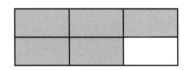 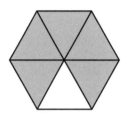

(c) If Sam read a book in 5 days, he must have read $\frac{1}{5}$ of the book each day. ☐

(d) When a 1-metre-long string is cut into 100 equal pieces, $\frac{1}{100}$ of it is 1 cm long. ☐

(e) The fewer parts a whole is equally divided into, the smaller each part becomes. ☐

5 Answer these multiple choice questions.

(a) A watermelon was cut into four equal pieces. Lewis ate three pieces. ☐ of the watermelon was left over.

A. $\frac{3}{4}$ B. $\frac{1}{2}$ C. $\frac{1}{4}$ D. $\frac{1}{3}$

(b) Count the stars in the picture. The number of ★ is ☐ of the whole.

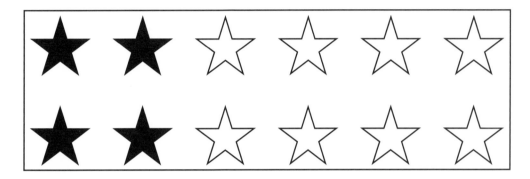

A. $\frac{1}{2}$ B. $\frac{1}{3}$ C. $\frac{3}{4}$ D. $\frac{1}{12}$

6 Six groups of pupils are planting trees. Each group needs to plant 8 trees. They have finished half of their target. How many more trees do they need to plant?

Answer: _____

_____

7 A class has 24 pupils with 6 girls and 18 boys.

(a) What fraction of the pupils in the class are girls, and what fraction are boys?

girls: _____   boys: _____

(b) If $\frac{1}{3}$ of the girls and half of the boys take part in the school play, what is the total number of the pupils participating inthe school play?

Answer: _____

Is the total number more or less than half of the number of the pupils in the class? Why?

Answer: _____

_____

_____

Chapter 7   Introduction to time (II)

7.1 Minute and hour

 Learning objective Solve problems involving equivalent times, using minutes, hours, days, weeks, months and years

 Basic questions

1 Look at the clock face and write the answers in the boxes.

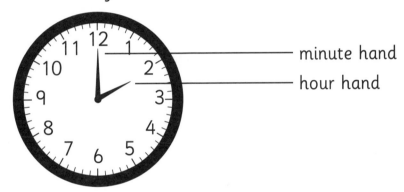

(a) There are ☐ hour marks and ☐ minute marks on the clock face.

(b) The clock shows the time of ☐ o'clock.

(c) When the hour hand moves from 1 to 2, we know that the time passed is ☐ hour.

2 Complete each sentence.

(a) There are ☐ minutes in an hour.

(b) There are ☐ hours in a day.

(c) There are ☐ days in five weeks.

(d) There are ☐ months in two years.

3 Which one is longer? Write >, < or = in each box.

(a) half an hour ☐ 30 minutes

(b) 50 minutes ☐ 1 hour

(c) 90 minutes ☐ 2 hours

(d) 2 days ☐ 100 hours

(e) quarter of an hour ☐ 20 minutes

(f) 5 days ☐ 75 hours

(g) 4 weeks ☐ 30 days

(h) half a year ☐ 7 months

(i) 65 minutes ☐ 1 hour

(j) 3 years ☐ 36 months

4 Saha played table tennis for half an hour and Meera played for 90 minutes. Are the following statements correct? Write a tick (✓) for yes and a cross (✗) for no.

(a) Saha played for 30 minutes. ☐

(b) Meera played for one hour and a half. ☐

(c) Saha played for half of the time Meera did. ☐

(d) Meera played for three times as long as Saha. ☐

5 Put the following measures of time in the correct order, starting with the shortest.

shortest

one minute

one hour

29 minutes

15 hours

15 days

half an hour

2 weeks

61 minutes

longest

7.2 Telling time to five minutes

Learning objective Read the time to the nearest five minutes

Basic questions

1 Look at the clock face and write the answers in the boxes.

(a) The time on the clock face shows it is ☐ o'clock.

(b) When the minute hand moves from 12 to 1, it shows that the time passed is ☐ minute(s).

(c) When it moves further from 1 to 2, it shows that another ☐ minute(s) has passed.

2 Look at the two different types of clocks. Draw lines to match the times.

  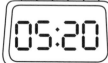 

07:15   11:05   05:20   03:45

3 Complete the table to show the time in words. Two have been done for you.

Time in numbers	Time in words
10:05	five past ten
9:55	
11:30	
3:50	
7:20	
1:45	a quarter to two
2:15	
10:00	
8:15	
6:45	

4 Read the clocks and write the time in words.

_____ past _____   _____ to _____

_____ past _____   _____ to _____

_____ past _____   _____ past _____

5 Draw the hour hand and minute hand to show the given time.

a quarter to five

twenty past two

twenty to two

a quarter past ten

Challenge and extension question

6 Write the names in order, starting with the person who finished the earliest.

Ali   Jaya   Mo   Jake   Amy

Ali: I finished the exercise at a quarter to three this afternoon.

Jaya: I finished it this morning.

Mo: I did not finish it until noon today.

Jake: I finished it at a quarter past three this afternoon.

Amy: I finished it yesterday.

Earliest

Latest

7.3 Comparing intervals of time

Learning objective Compare and sequence intervals of time, including minutes, hours and days

Basic questions

1 Look at the sets of clocks. How much time has passed? Write your answers in words.

(a)

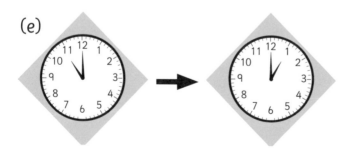

(b)

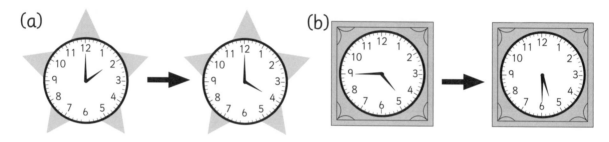

(c) 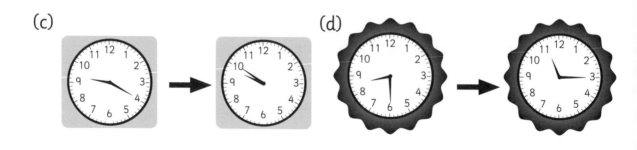 (d)

(e)

2 Mr Scott visited Ghana from 15 May to 31 May. The duration of his visit was ☐ days, or ☐ weeks and ☐ days.

3 Complete the table.

Half an hour ago	Now	An hour later
	7 o'clock	
half past 10		
		9 o'clock
	half past 4	

4 True or false? Put a tick (✓) for true or a cross (✗) for false in each box.

(a) There are 72 hours in 3 days. ☐

(b) Hema watches a TV news programme from half past six to seven o'clock in the evening. This is one hour and 30 minutes. ☐

(c) The shop opens from half past eight in the morning and closes at half past five in the afternoon. This is 9 hours trading time. ☐

(d) If a three-day celebration starts on Monday, it will end on Friday. ☐

Challenge and extension question

5 Write the names in order, starting with the person who finished the task in the shortest time.

Ali   Jaya Mo   Jake Amy

Ali: I started at one in the afternoon and finished at a quarter to three.

Jaya: I started at half past eight and finished at ten to ten.

Mo: I started at 12 noon yesterday and finished at a quarter past nine today.

Jake: I started at five past one in the afternoon and finished at a quarter past three.

Amy: I started at 8:00 in the evening yesterday and finished at 10:00 on the same day.

Shortest

Longest

Chapter 7 test

1 Look at the different kinds of clocks. Draw lines to match the times.

2 True or false? Put a tick (✓) for true or a cross (✗) for false.

(a) There are 60 minutes in an hour and 24 hours in a day. ☐

(b) $\frac{1}{4}$ of an hour is 30 minutes. ☐

(c) $\frac{1}{3}$ of half an hour is 10 minutes. ☐

(d) Ben started reading a book at a quarter to nine and stopped at twenty to ten. He took less than one hour to read it. ☐

(e) Emily started a drawing at eight o'clock in the morning and finished it at a quarter to eleven. She took more than three hours to complete her drawing. ☐

3 Draw the hour hand and a minute hand on the clock faces below to show the given time.

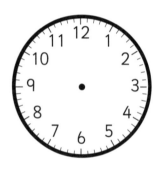

seven o'clock

five to five

a quarter past one

twenty-five past eleven

4 Write the times in the table below.

One hour earlier	Now	A quarter of an hour later
	four o'clock	
ten to three		
	a quarter past eleven	
		half past six
a quarter to eight		

5 Which is the longer time? Write >, < or = in each circle.

(a) one hour ◯ 55 minutes

(b) half an hour ◯ 25 minutes

(c) a quarter of an hour ◯ 15 minutes

(d) 3 weeks ◯ 22 days

(e) three quarters of an hour ◯ 50 minutes

(f) 4 years ◯ 48 months

6 A school organised a trip for pupils to visit a theme park.

1. Leaving school ...

2. Arriving at the theme park ...

3. Leaving the theme park ...

4. Arriving back at school ...

(a) How long did it take to get from school to the theme park?

_____

(b) How many hours did the pupils spend in the theme park?

_____

(c) How long did it take to get from the theme park back to school?

_____

(d) How many hours did the whole trip take?

_____

Chapter 8　Mass, temperature and capacity

8.1　Light and heavy (1)

 Learning objective Compare the mass of objects using a balance

 Basic questions

1 Read and complete.

Seesaws can be used to compare the masses of objects.

If the seesaw is balanced, it means the masses of the objects on both sides are the same.

If one side is lower, it means the object on that side is _____.

2 Let's compare. Which one is heavier, and which is lighter?

(a) An egg and a bottle of juice

(b) A schoolbag and a pencil case

_____ is heavier than

_____.

_____ is lighter than

_____.

(c) A watermelon and
an apple

_____ is lighter

than _____ .

(d) A mobile phone and
a computer

_____ is heavier

than _____ .

3 Look at the pictures and compare the masses of the objects.
Circle the lighter objects.

4 These pieces of fruit are hanging from identical rubber bands.
Complete the table to show the pieces of fruit in order of their
masses from the heaviest to the lightest.

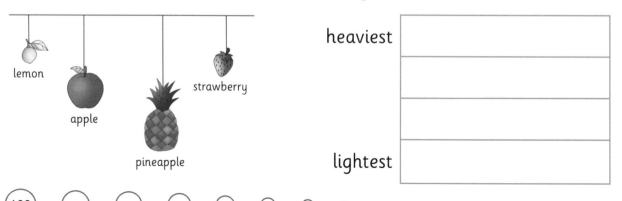

lemon

apple

pineapple

strawberry

heaviest	
lightest	

Challenge and extension question

5 Compare the weights of animals.

Put a tick (✓) in the circle for the heavier and a cross (✗) for the lighter of each pair.

giraffe ◯ panda ◯   rabbit ◯ panda ◯

puppy ◯ rabbit ◯

6 How many cubes are needed to keep the scales balanced?

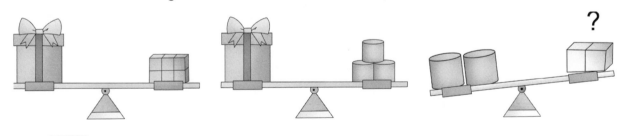

☐ small cubes are needed to keep the scales balanced.

8.2 Light and heavy (2)

 Learning objective Compare the masses of objects using a balance

 Basic questions

1 The masses of the following objects are measured in terms of small cubes.

Objects	Maths book	Pencil case	Pencil sharpener	Mouse	Mug
Number of cubes	12	8	2	5	10

Draw lines to match the objects from the lightest to the heaviest.

| Maths book | Pencil case | Pencil sharpener | Mouse | Mug |

(lightest)   (heaviest)

2 Write numbers in the boxes to put the following fruit in order from the lightest (1) to the heaviest (5).

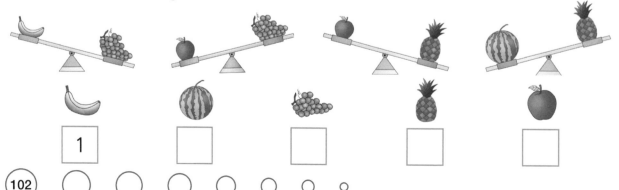

1

3 The cat is heavier than the cockerel.

Write numbers in the boxes to put the following animals in order from the lightest (1) to the heaviest (4).

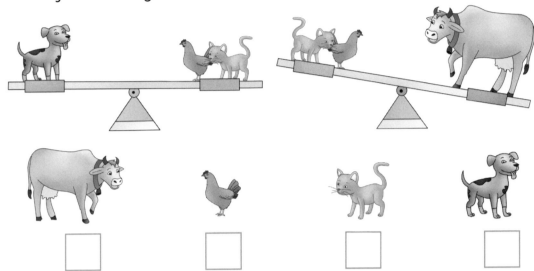

☐   ☐   ☐   ☐

4 Write numbers in the boxes to put the following fruits in order from the lightest (1) to the heaviest (4).

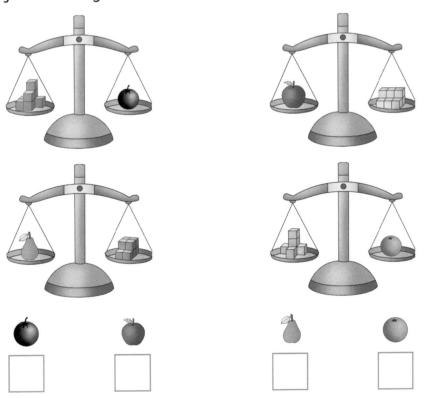

☐   ☐   ☐   ☐

5 The pupil wearing number ☐ is the heaviest. The pupil wearing number ☐ is the lightest.

Challenge and extension questions

6 (a)

△ = _____ ○

(b)

▢ = _____ △

7 The mass of 1 watermelon = the mass of 3 pineapples.

The mass of 1 pineapple = the mass of 3 apples.

(a) The mass of 1 watermelon = the mass of ☐ apples.

(b) The mass of 18 apples = the mass of ☐ watermelons.

8.3 Knowing grams and kilograms

Learning objective Compare grams and kilograms and use appropriate units to measure mass

Basic questions

1 Read the scale to find out the mass. Circle the suitable unit (grams or kilograms).

| 3 | 98 | 68 | 2 |

| g | kg | g | kg | g | kg | g | kg |

2 Write grams or kilograms into each space.

The mass of a pack of flour is 1 _____.

Ella's body mass is 28 _____.

A carton of milk weighs $\frac{1}{2}$ _____.

The mass of a hen is 2 _____.

The mass of an egg is 70 _____.

The mass of a bag of rice is 5 _____.

The mass of a cow is 100 _____.

A bottle of medicine weighs 100 _____.

The mass of a cat is 3 _____.

The mass of a table tennis ball is 3 _____.

3 Multiple choice questions.

(a) The correct order from the heaviest to the lightest is [   ] .

A. Pencil case, doll, mug   B. Mug, doll, pencil case

C. Doll, pencil case, mug   D. Mug, pencil case, doll

(b) The abbreviation for the mass unit 'gram' is [   ].

A. m   B. km   C. g   D. kg

(c) Compare the masses of 3 kilograms of iron and 3 kilograms of cotton. Which statement is true? [   ]

A. The iron is heavier.

B. The cotton is heavier.

C. The iron is as heavy as the cotton.

D. They are not comparable.

(d) Compare the measurements, 5 metres and 5 kilograms.

Which statement is true? ☐

A. 5 metres is bigger.   B. 5 kilograms is bigger.

C. They are the same.   D. They are not comparable.

Challenge and extension question

4 Kim used this method to find out the mass of the kitten.
Can you work out the mass of the kitten?

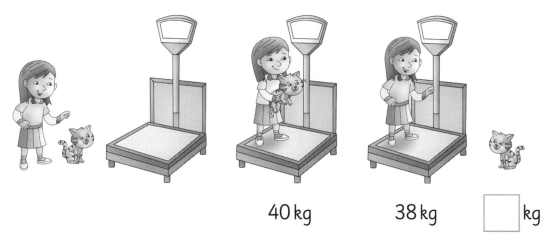

40 kg   38 kg   ☐ kg

Write a number sentence to show your answer.

_____

8.4 Reading scales

Learning objective Use scales to measure mass

Basic questions

1 Write 'g' or 'kg' to show which objects would be weighed in grams and which would be weighed in kilograms.

   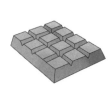 

_____ _____ _____ _____ _____

2 Draw lines to match the pictures to their descriptions.

Digital weighing scales

Analogue weighing scales

Balance

3 Write the mass shown on each weighing scales. Remember to write the abbreviation kg to show these are kilograms.

_____ _____ _____

_____ _____

4 Draw the needle on the display on the scales below to show the mass of the strawberries.

_____ _____ _____ _____

Challenge and extension question

5 What is the smallest number of objects that could be put on the scale to make the display show 15 kg?

2.5 kg

1.5 kg

5 kg

4.5 kg

1 kg

2 kg

3 kg

8.5 Temperature

Learning objective Estimate and measure temperatures using a thermometer

Basic questions

1 Read the temperature shown on the thermometer and fill in each box below.

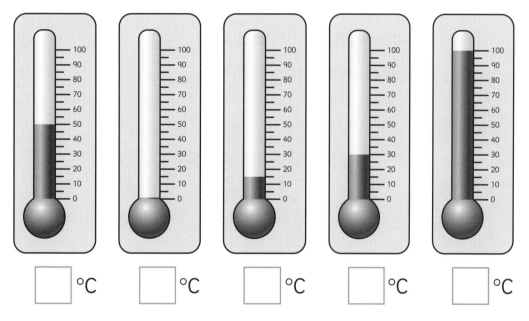

☐ °C   ☐ °C   ☐ °C   ☐ °C   ☐ °C

2 True or false? Put a tick (✓) for true or a cross (✗) for false in each box.

(a) Temperature tells us how hot or cold an object or a place is. ☐

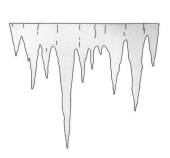

(b) The temperature in the daytime is usually lower than the temperature at night-time. ☐

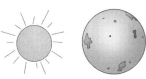

(c) The higher the temperature in a place, the warmer people will feel. ☐

(d) Summer is usually the hottest season, with the highest daily temperature in the year. ☐

3 Draw a line to connect each matching pair.

  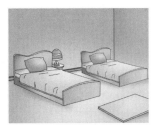 

| Boiling water | Body temperature | Hotel room temperature | Water freezing temperature |

37 °C   0 °C   100 °C   25 °C

4 The table below shows the temperature of a city taken at noon on the first day of each month. Fill in the blanks based on the table.

Month	Jan	Feb	Mar	Apr	May	Jun	Jul	Aug	Sep	Oct	Nov	Dec
Temperature (°C)	6	8	9	11	14	16	19	19	17	13	10	7

(a) What was the temperature of the city in April?

_____

(b) Which two months had the same temperature on the first day?

_____

(c) Which month had the coldest first day? What was the temperature in that month?

_____

Challenge and extension question

5 Use a thermometer and record the temperature in your school. Fill in the blanks. Put only whole numbers in your recording.

(a) In your classroom.

Temperature: _____

Date/time of measurement: _____

(b) In your playground.

Temperature: _____

Date/time of measurement: _____

(c) It is warmer in the _____.

Date: _____

8.6 Introduction to capacity and volume

Learning objective Compare millilitres and litres and use appropriate units to measure capacity

Basic questions

1 (a) Draw lines to match the glasses of water to their descriptions.

| Full | Empty | Less than half full | Half full | More than half full |

(b) Colour the glass that holds the largest volume of water.

(c) Draw a circle around the glass that has the smallest volume of water.

2 Look at these boxes of different capacity.

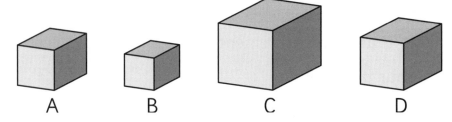

A   B   C   D

Put the boxes in order from the least capacity to the greatest capacity.

3 Which of the following has the greater capacity, and by how much?

Put a tick (✓) for the greater and a cross (✗) for the smaller.

(a)

☐   ☐

The difference in the capacities is _____.

(b)

☐   ☐

The difference in the capacities is _____.

4 Fill in the blanks with 'litres' (l) or 'millilitres' (ml).

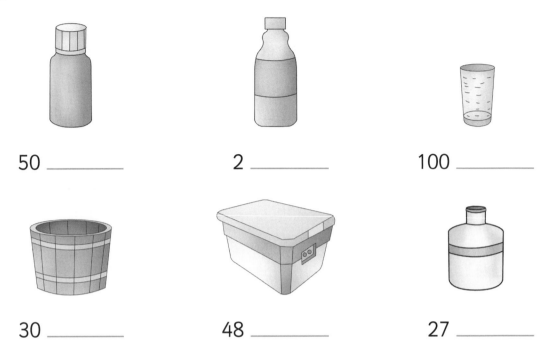

50 _____   2 _____   100 _____

30 _____   48 _____   27 _____

Challenge and extension question

5 Try this yourself. Use a suitable measuring tool to find out the capacity of an object. For example, you might choose a small cup to find the capacity of a bottle or kettle.

Record your results below.

Measuring tool used: _____

Object measured: _____

Capacity of the object: _____

Date of measurement: _____

8.7 Capacity and volume

 Learning objective Use scales to measure capacity and volume

Basic questions

1 Look at the containers below. Do you think it would be better to measure the capacity of each in millilitres or litres? Write ml or l under each container to show which units would be better to measure its capacity.

_____ _____ _____ _____ _____ _____

2 Five identical glasses have been filled with water. Order them 1 to 5, with 1 for the glass with the least volume and 5 for the glass with the greatest volume.

_____   _____   _____   _____   _____

3 The table shows the volumes of four of the items pictured below. Can you work out which ones? Complete the table by writing in the correct items. Remember: two of the items pictured are not included in the table.

Volume	Name of item
54 litres	
2 litres	
19 litres	
10 litres	

4 (a) What is the total volume of juice in these measuring cylinders?

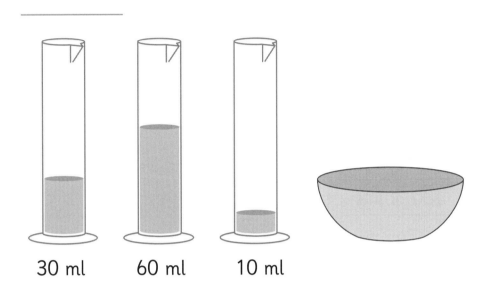

30 ml   60 ml   10 ml

(b) How could you use these measuring cylinders to pour 70 ml into the bowl?

List the steps you would take:

First, _____

Then, _____

_____

_____

_____

(c) Can you think of another way? If so, describe it.

_____

_____

Chapter 8 test

1 Put the following in order from the heaviest to the lightest.
(Note: 100 grams < 1 kilogram)

Heaviest

2 kilograms

3 grams

80 grams

25 kilograms

100 grams

Lightest

2 To measure the mass of an object, you can put the object on one side of a balance scales, and small identical cubes on the other side of the scales. Answer the questions based on the results shown in the table.

(a) The heaviest object
is _____.
The lightest object
is _____.

(b) The pen is _____ cubes heavier than the ruler.

(c) The total mass of the pen and the rubber is the same as the total mass of the _____ and the _____.

Objects	Number of cubes
ruler	2
stapler	11
pen	10
rubber	3

3 Fill in the boxes with suitable units of measurement.

   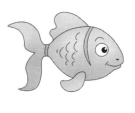

Weight: 50 ☐ Weight: 1 ☐ Weight: 80 ☐ Weight: 20 ☐

  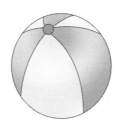 

Weight: 23 ☐ Weight: 3 ☐ Weight: 15 ☐ Weight: 100 ☐

4 Draw lines to match the objects to the measurements below.

| 25 °C | 5 millilitres | 12 kg | 40 litres | 80 g | 1 °C |

5 Application problems.

(a) The mass of a packet of sweets is 20 grams. What is the mass of 5 packets? Write your answer in grams.

Answer: _____

(b) The table shows the temperature of a city taken at noon on the warmest and coldest day of each season. Use the table to answer the questions.

	Winter	Spring	Summer	Autumn
Low temperature	6°C	9°C	18°C	10°C
High temperature	8°C	15°C	21°C	18°C

(i) What was the low temperature in summer?

_____

(ii) What was the high temperature in winter?

_____

(iii) Which season had the greatest difference between the low and high temperatures?

_____

(iv) Which season had the least difference?

_____

6 Calculate with reasoning.

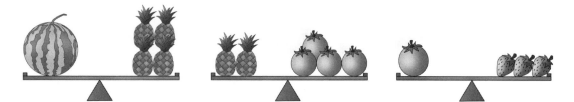

Weight: 1 🍉 = ☐ 🍓

7 Conor used a 5 ml measuring beaker to measure the capacities of a plastic cup and a paper cup.

The plastic cup was filled up to full capacity with 3 beakers of water to the 5 ml mark.

The paper cup was filled up with 6 beakers of water to the 5 ml mark.

Answer the following questions.

(a) Which cup has greater capacity? _____

(b) How many times greater is the capacity of the paper cup than the plastic one?

_____

(c) What is the total capacity of these two cups? _____

Chapter 9 Let's practise geometry

9.1 Position, turn and direction

Learning objective Use clockwise and anticlockwise turns to describe movement

Basic questions

1 The figures below show different directions of rotation. Write 'clockwise rotation' or 'anticlockwise rotation' in the spaces.

(a)

The clock face on the left shows

_____.

The clock face on the right shows _____.

(b)

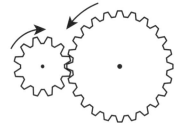

The circle inside shows

_____.

The circle outside shows

_____.

(c)

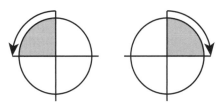

The smaller gear wheel shows

_____.

The bigger gear wheel shows

_____.

(d)

The circle on the left shows

_____.

The circle on the right shows

_____.

2 True or false? Put a tick (✓) for true or a cross (✗) for false in each box.

(a) The hour hand of a clock rotates clockwise.

(b) A right angle represents a quarter turn.

(c) Two right angles make a complete turn.

(d) Starting from the same position, a half turn clockwise and a half turn anticlockwise will result in the same ending position.

(e) If you start off facing East, after you make a three-quarter turn anticlockwise, you will face North.

3 The pictures below show a person in different positions. Fill in the spaces in terms of turns clockwise or anticlockwise.

Can you act it out by yourself?

Position A   Position B   Position C   Position D

(a) From Position A to Position B, a _____ turn anticlockwise is needed.

(b) From Position A to Position C, a _____ turn clockwise is needed.

(c) From Position A to Position C, a _____ turn anticlockwise is needed.

(d) From Position B to Position C, a _____ turn clockwise is needed.

(e) From Position C to Position D, a _____ turn clockwise is needed.

4 Kevin the kitten can't decide what he wants to do. He keeps turning around to face his favourite things.

How far does Kevin turn to face his favourite things? The first one has been done for you.

(a) Kevin is facing his bed; he makes a quarter turn clockwise to face his ball of wool.

(b) Kevin is facing the garden; he makes a _____ clockwise to face his bed.

(c) Kevin is facing his food; he makes a _____ clockwise to face the garden.

(d) Kevin is facing his ball of wool; he makes a a three-quarter turn _____ to face his food.

(e) Kevin is facing his ball of wool; he makes a quarter turn _____ to face his bed.

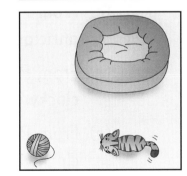

5 Describe how a robot can move through the square maze from the entrance position M to the exit position N in the shortest route. The first two steps have been done for you.

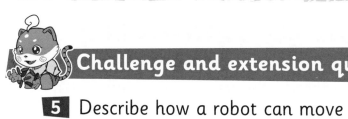

Step ①: move one square in a a straight line to position A.

Step ②: make a quarter turn clockwise and then move 4 squares in a straight line to position B.

Step ③: _____

_____

Step ④: _____

_____

Step ⑤: _____

_____

Step ⑥: _____

_____

Step ⑦: _____

_____

Step ⑧: _____

_____

Step ⑨: _____

_____

9.2 Introduction to symmetry

Learning objective Identify the vertical line symmetry of shapes

Basic questions

1 Does each of the following pictures have symmetry? Put a tick (✓) for yes or a cross (✗) for no in each box.

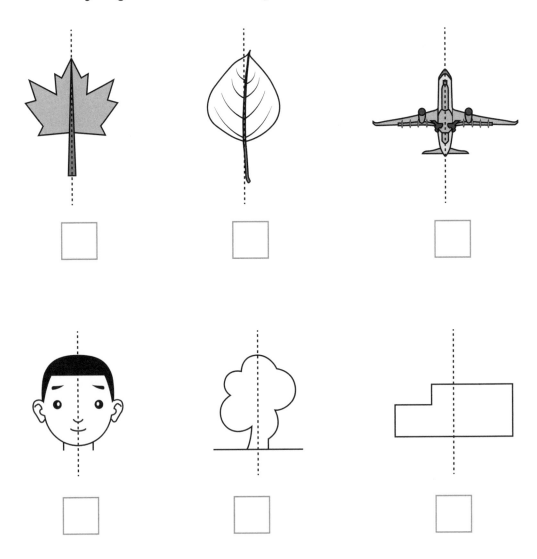

2 Use a straight edge to draw a line to divide each figure into two identical halves.

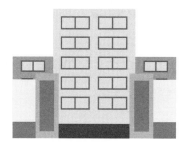

Challenge and extension question

3 Each picture below shows half of a symmetrical figure. Draw the other half.

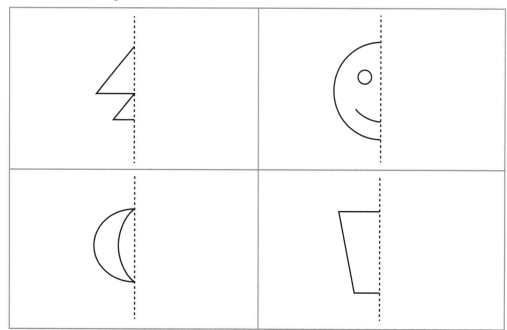

9.3 Simple properties of 2-dimensional shapes

Learning objective Identify and describe simple properties of 2-D shapes

Basic questions

1 Draw a line to match each pair.

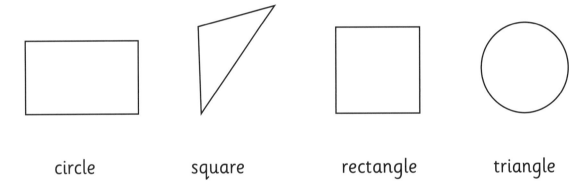

circle   square   rectangle   triangle

2 How many sides and vertices are there in each shape below?

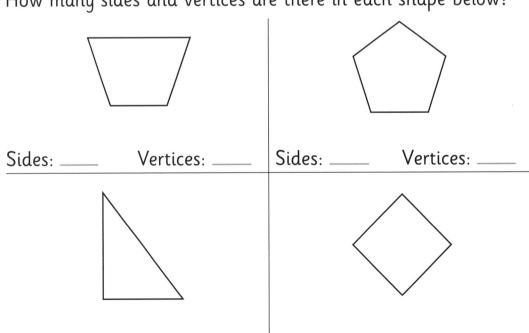

Sides: ____   Vertices: ____   Sides: ____   Vertices: ____

Sides: ____   Vertices: ____   Sides: ____   Vertices: ____

3 Use a straight edge to draw a line to divide each figure into two identical halves.

4 True or false? Put a tick (✓) for true or a cross (✗) for false in each box.

(a) A square has as many sides and vertices as a rectangle. ☐

(b) Any triangle must be symmetrical. ☐

(c) Any square must be symmetrical. ☐

(d) Both squares and rectangles are quadrilaterals. ☐

(e) A polygon must have at least four sides. ☐

(f) A circle is not a polygon. ☐

(g) A quadrilateral has more sides and vertices than a triangle. ☐

Challenge and extension questions

5 The figure below shows a rectangle.

(a) Does it have symmetry? _____

(b) How many sides and vertices does it have?

_____

(c) If a corner of the rectangle is cut off, how many sides will the remaining shape have? Draw to show your answer.

6 Draw a polygon that has eight sides. How many vertices are there?

9.4 Simple properties of 3-dimensional shapes

Learning objective Identify and describe simple properties of 3-D shapes

 Basic questions

1 Write the name for each 3-dimensional shape below.

_____   _____   _____

 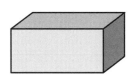

_____   _____

2 Complete the table.

Figure		
Name	cube	cuboid
Number of edges		
Number of vertices		
Number of faces		

3 Complete the shape facts.

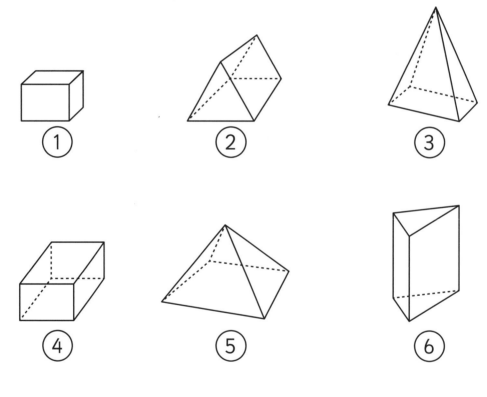

The pyramid has ☐ edges, ☐ vertices and ☐ faces.

The prism has ☐ edges, ☐ vertices and ☐ faces.

4 Look at the pictures. Write the numbers of the shape onto the lines to answer the questions below.

①　②　③

④　⑤　⑥

(a) Shape ____ and Shape ____ each have 8 vertices.

How many edges and faces do they each have?

Edges: _____

Faces: _____

(b) Shape ____ and Shape ____ each have 8 edges.

How many vertices and faces do they each have?

Vertices: _____

Faces: _____

(c) Shape ____ and Shape ____ each have 9 edges.

How many vertices and faces do they each have?

Vertices: _____

Faces: _____

Challenge and extension question

5 How many edges, vertices and faces does each shape have?

Edges: ____

Vertices: ____

Faces: ____

Edges: ____

Vertices: ____

Faces: ____

Edges: ____

Vertices: ____

Faces: ____

9.5 Relating 2-D shapes and 3-D shapes

Learning objective Identify 2-D shapes as the faces of 3-D shapes

Basic questions

1 Sort the shapes by writing the number of each shape into the correct box.

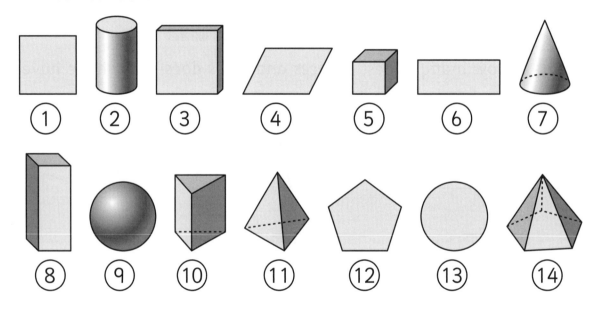

2-dimensional shapes	3-dimensional shapes

2 Draw a line to link each 3-D shape with the correct faces label.

pyramid

cube

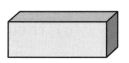
cuboid

All faces are rectangles	All faces are triangles	All faces are squares

3 Count the number of different 2-dimensional faces that each of these 3-D figures has.

Write 0 if a 3-D shape does not have the shape of face indicated.

square base

Square face(s): ____

Rectangular face(s): ____

Triangular face(s): ____

Circular face(s): ____

Square face(s): ____

Rectangular face(s): ____

Triangular face(s): ____

Circular face(s): ____

Square face(s): ____

Rectangular face(s): ____

Triangular face(s): ____

Circular face(s): ____

4 Do both pyramids and prisms have the following properties? Put a tick (✓) for yes or a cross (✗) for no in each box.

(a) They have only flat faces. ☐

(b) All the faces must be triangles. ☐

(c) All the faces must be quadrilaterals. ☐

(d) There are no circular faces. ☐

Challenge and extension question

5 Draw each of the following 3-D shapes as described.

(a) It has exactly 6 square faces.

(b) It has exactly 4 triangular faces.

(c) It has exactly 1 quadrilateral face and 4 triangular faces.

9.6 Ordering combinations of shapes and objects

Learning objective Order and arrange mathematical objects in patterns and sequences

Basic questions

1 Write the shape letters below to order these 2-D shapes, from smallest to greatest number of sides.

Shape A

Shape B

Shape C

Shape D

Shape E

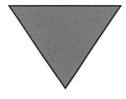

Shape F

Answer: _____

Smallest number of sides   Greatest number of sides

$\longrightarrow$

2 (a) Complete these sequences.

| △ | ○ | ✚ | △ | | | | | |

| ✖ | ☺ | ♡ | ◺ | ✖ | | | | |

| ⬆ | ➡ | ⬇ | ⬅ | | | | | |

(b) Now make your own sequence.

| | | | | | | | | |

3 Order these 3-D shapes by the number of faces that they have.

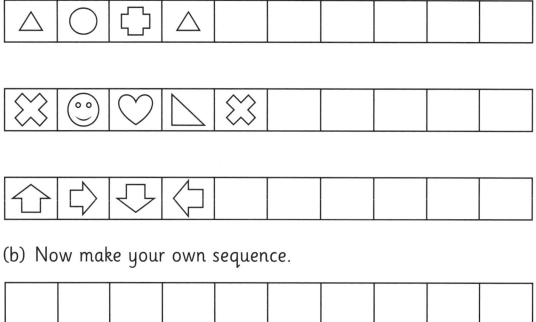

Shape A   Shape B   Shape C

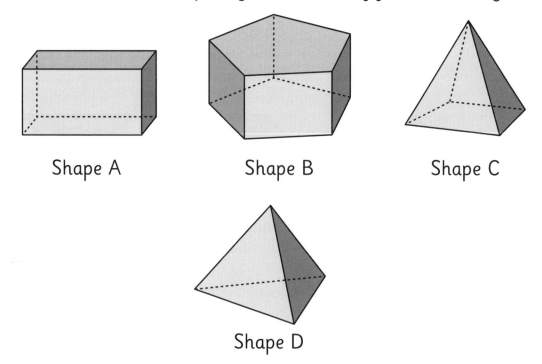

Shape D

The shape with the most faces is Shape _____ followed by

Shape _____ . Next, it is Shape _____ and finally Shape _____ .

Challenge and extension question

4 What would the 10th shape be in this sequence?

Answer: _____

Explain how you know.

_____

_____

_____

1 Look at the diagrams. Write the number of each shape on the correct line.

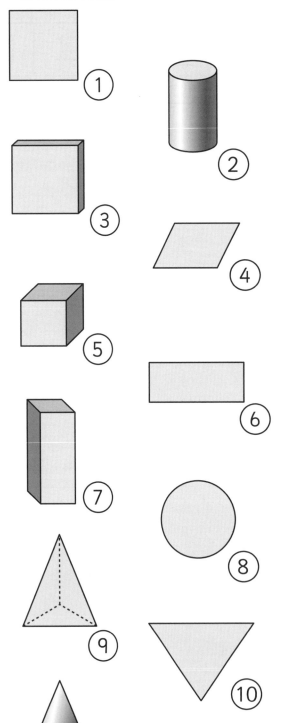

cone _____

pyramid _____

cube _____

cuboid _____

circle _____

triangle _____

square _____

rectangle _____

2 True or false? Put a tick (✓) for true or a cross (✗) for false in each box.

(a) Two right angles make a half turn. □

(b) A prism has six quadrilateral faces. □

(c) All the faces of a pyramid are triangles. □

(d) Both a square and a rectangle have four sides and four vertices. □

(e) Both a rectangle and a square have symmetry. □

(f) A polygon must have at least three sides. □

(g) A circle is symmetrical. □

(h) A cone has two circular faces. □

3 Do the following shapes have symmetry?

Put a tick (✓) for yes or a cross (✗) for no. If the answer is yes, draw a line to divide the shape into two identical halves.

□   □   □

4 A robot starts in the upper-right corner in the 10 by 10 square grid, as shown below. It makes the following moves:

Move 1: 5 squares downwards in a straight line.

Move 2: A quarter turn clockwise and then a move 4 squares in a straight line.

Move 3: A three-quarter turn anticlockwise and then a move 5 squares in a straight line.

Move 4: A quarter turn anticlockwise and then a move 5 squares in a straight line.

Draw lines or arrows to show the robot's movement across the grid. What is the final position of the robot?

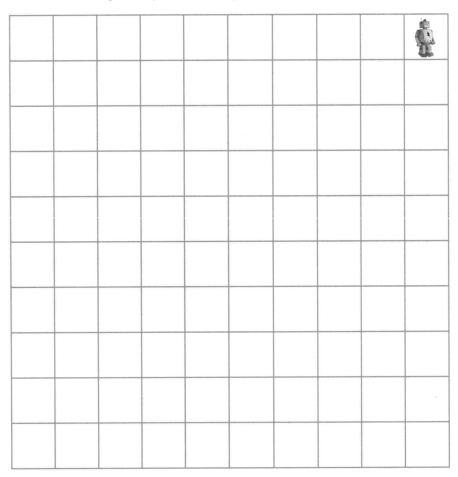

Answer: The final position of the robot is

5 Look at the 3-dimensional shapes below and write the number of each item indicated. (Write 0 if it is not applicable.)

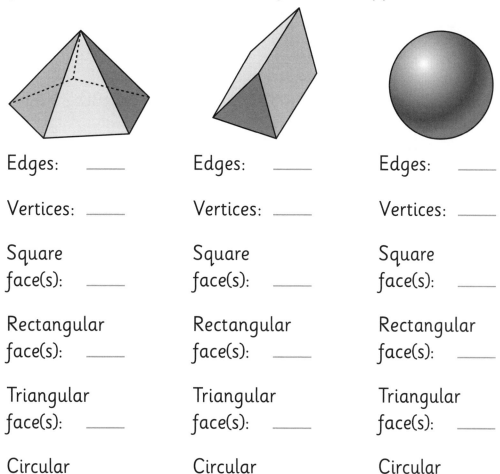

| Edges: _____ | Edges: _____ | Edges: _____ |

| Vertices: _____ | Vertices: _____ | Vertices: _____ |

| Square face(s): _____ | Square face(s): _____ | Square face(s): _____ |

| Rectangular face(s): _____ | Rectangular face(s): _____ | Rectangular face(s): _____ |

| Triangular face(s): _____ | Triangular face(s): _____ | Triangular face(s): _____ |

| Circular face(s): _____ | Circular face(s): _____ | Circular face(s): _____ |

6 Draw a 3-dimensional shape that has one quadrilateral face and four triangular faces.

End of year test

1 Calculate mentally. (12%)

5 × 5 = ☐   24 ÷ 4 = ☐   56 ÷ 8 = ☐

72 ÷ 9 = ☐   9 + 68 = ☐   90 − 79 = ☐

6 × 10 = ☐   10 × 10 − 70 = ☐   5 + 5 × 7 = ☐

65 − 50 = ☐   12 ÷ 6 = 12 − ☐   6 ÷ 2 = ☐ ÷ 6

2 Count in fractions of $\frac{1}{2}$ and $\frac{1}{4}$ using the number lines and fill in the boxes. (6 %)

(a)
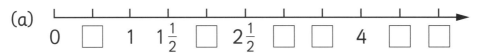

(b)
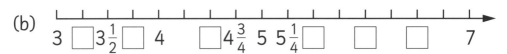

3 Draw a line to match each pair. (4%)

Mass and weight	Hot or cold
Length	Space taken up
Capacity and volume	Heavy or light
Temperature	Long or short

4 Fill in the brackets with suitable units of measurement. (8%)

A lesson lasts 35 (   ).   The movie lasted 2 (   ).

An adult's body mass is 85 (   ).

The mass of an eraser is 11 (   ).

The mass of a ball pen is 6 (   ).

The mass of a bag of fruits is 6 (   ).

The capacity of a tea cup is 150 (   ).

The temperature of boiling water is 100 (   ).

5 Put the following in order. (4%)

(a) from the lightest to the heaviest in terms of mass.

1 kilogram, 20 grams, 20 kilograms, 40 kilograms, 88 grams, 100 kilograms, 100 grams

_____

_____

_____

(b) from the shortest to the longest in terms of time.

2 days, half an hour, 2 weeks, 100 hours, a quarter of an hour, 20 minutes

_____

_____

_____

6 Think carefully and then answer. (8%)

(a) Write >, < or = into each ◯.

6×8 ◯ 42 7×6 ◯ $6 + 7$

2×2 ◯ $2 + 2$ $80 \div 10$ ◯ 4×4

$0 + 10$ ◯ 0×10

(b) What is the greatest number you can fill in the ( )?

$3 \times (\quad) < 22$ $4 \times (\quad) < 7 \times 6$ $(\quad) - 6 < 9 \times 3$

7 Write number sentences for the following. (12%)

(a) Write four number sentences whose products are all 24.

_____ _____ _____ _____

(b) Write four number sentences whose quotients are all 5.

_____ _____ _____ _____

(c) Write four multiplication or division sentences whose answers can be obtained using the same multiplication fact.

_____ _____ _____ _____

8 Application problems. (20%)

(a) Make a count first. How many animals of different kinds are there? Then complete the tally chart and the block diagram.

Tally chart	
Animal	Number

Block diagram

10			
9			
8			
7			
6			
5			
4			
3			
2			
1			

(i) Use the block diagram to answer the following.

(ii) There are ( ) animals altogether.

(iii) There are ( ) times as many as .

(b)

Each plate contains (　) pears. There are (　) plates.

There are (　) pears altogether.

☐ ◯ ☐ = ☐

There are 15 pears.

(　) pears are put on each plate. There are (　) plates.

☐ ◯ ☐ = ☐

Put 15 pears into (　) plates equally.

Each plate has (　) pears.

☐ ◯ ☐ = ☐

(c) The pictogram below shows the number of trees that three groups of pupils planted in a week.

Group 1	Group 2	Group 3
△△△△	△	△△△△△

Each △ stands for 3 trees.

(i) Which group planted the most trees? (　)

(ii) Write the number of trees Group 2 planted as a fraction of the number of trees Group 1 planted. (　)

(iii) How many trees did all three groups plant in total? (　)

(d) Write number sentences and answer each question.

Dad   Mo   Grandpa

Grandpa is 63 years old. Mo is 7 years old. Dad is 5 times as old as Mo.

How old is Dad?

Number sentence: _____

Answer: Dad is (   ) years old.

How many years older is Grandpa than Dad?

Number sentence: _____

Answer: Grandpa is (   ) years older than Dad.

How many times is Grandpa as old as Mo?

Number sentence: _____

Answer: Grandpa is (   ) times as old as Mo.

9 Multiple choice questions. (18%)

(a) Which of the following is incorrect? (   )

    A. A rectangle has one more side than a triangle.

    B. Both squares and rectangles have symmetry.

    C. A pyramid must have four trianglular faces and one quadrilateral face.

(b) The number sentence is 8 ÷ 4 = 2.

Which of the following expressions is incorrect? (   )

A. What is 4 times 8?

B. How many times 4 is 8?

C. Divide 8 units into 4 pieces equally. How many units are there in each piece?

(c) The sum of the following can be represented using the number sentence (   ).

A. 5 × 4 + 2   B. 6 × 4   C. 6 × 4 + 2

(d) Kaya cut a loaf of bread into 3 pieces equally. She ate 2 pieces. (   ) of the bread was left over.

A. $\frac{3}{4}$ B. $\frac{1}{3}$ C. $\frac{1}{4}$

(e) 2 + 2 can be written as 2 × 2, and 3 + 3 can be written as (   ).

A. 3 × 3   B. 2 × 3   C. 2 + 3

(f) Two right angles make (   )?

A. a quarter turn   B. a half turn   C. a complete turn

(g) Scott was facing North at first and then made a three-quarter turn clockwise. He faces (   ) at the end.

A. South   B. West   C. East

(h) When Ben finished his homework in the evening, he looked at the clock as shown on the right. He started his homework at a quarter past seven. He spent (   ).

A. 65 minutes   B. 25 minutes   C. 35 minutes

(i) The difference between the greatest two-digit number and the least two-digit number is (   ).

A. 89   B. 90   C. 99

10 Look at the 3-dimensional figures below and write the number of each item. (8%)

Edges: _____   Edges: _____   Edges: _____

Vertices: _____   Vertices: _____   Vertices: _____

Square
face(s): _____   Square
face(s): _____   Square
face(s): _____

Rectangular
face(s): _____   Rectangular
face(s): _____   Rectangular
face(s): _____

Triangular
face(s): _____   Triangular
face(s): _____   Triangular
face(s): _____

Circular
face(s): _____   Circular
face(s): _____   Circular
face(s): _____

Notes

Notes

Notes

Notes